DIY OFF GRID SOLAR PANEL FOR BEGINNERS

UNLOCK ENERGY FREEDOM IN 4 SIMPLE STEPS | BUDGET FRIENDLY SOLAR POWER INSTALLATION GUIDE

CHRISTOPHER GARCIA

Copyright © 2024 by Christopher Garcia

All rights reserved. No part of this publication may be reproduced, distributed, or transmitted in any form or by any means, including photocopying, recording, or other electronic or mechanical methods, without the prior written permission of the publisher, except in the case of brief quotations embodied in critical reviews and certain other noncommercial uses permitted by copyright law. For permission requests, write to the publisher, addressed "Attention: Permissions Coordinator," at the address below.

TABLE OF CONTENTS

INTRODUCTION TO ENERGY INDEPENDENCE 11
- The Rising Popularity of Solar Power 11

CHAPTER 1 15
SOLAR POWER FUNDAMENTALS 15
- The Science of Solar Energy 16
- Components of a Solar Power System 17
- Off-Grid vs. On-Grid 19
- Advantages of Off-Grid Living 20

CHAPTER 2 25
INITIAL CONSIDERATIONS FOR GOING SOLAR 25
- Assessing Your Energy Needs 25
- Solar Potential and Climate Considerations 27
- Budgeting for Your Solar Project 30
- Understanding Your Site's Solar Access 31

CHAPTER 3 33
NAVIGATING REGULATIONS AND INCENTIVES 33
- Local Zoning and Permitting Process 33
- Understanding Zoning Laws 33
- Federal and State Solar Incentives 35
- How to Claim Solar Tax Credits 37
- Net Metering and Other Incentives 39

CHAPTER 4 43
STEP 1 DESIGNING YOUR SYSTEM 43
- Determining System Size and Capacity 43
- Selecting the Right Solar Panels 44
- Charge Controllers PWM vs. MPPT 46
- Battery Storage Options 48
- Choosing an Inverter for Your Needs 49
- Balancing System Efficiency and Cost 52

CHAPTER 5 55
STEP 2 PROCURING COMPONENTS 55
- Where to Buy Solar Equipment 57
- Evaluating Component Quality 58
- Tips for Budget Friendly Purchases 60
- DIY vs. Professional Installation Components 62

CHAPTER 6 65
STEP 3 INSTALLATION PROCESS 65
- Safety and Preparation 65
- Mounting Solar Panels Techniques and Tips 66
- Electrical Wiring Basics 67
- Installing the Charge Controller and Inverter 69
- Setting Up the Battery Bank 71
- System Grounding and Protection 73

CHAPTER 7 75
STEP 4 TESTING AND MAINTENANCE 75
- Initial System Testing and Troubleshooting 75
- Routine System Maintenance 76
- Monitoring and Optimizing Performance 78
- Upgrades and System Expansion 80

CHAPTER 8 83
LIVING OFF THE GRID 83
- Daily Life with Solar Power 84
- Energy Conservation Tips 86
- Dealing with Power Outages 88
- Long-Term Sustainability 90

CHAPTER 9 95
BEYOND THE BASICS 95
- Hybrid Systems Incorporating Wind and Hydro 97
- Solar-Powered Water Heating 99
- Automating Your Solar Power System 100
- Community Solar Projects and Sharing 102

CHAPTER 10 105
OTHER RESOURCES 105
- Installation and Maintenance Checklists 107
- Solar Power System Troubleshooting: A comprehensive guide on diagnosing and fixing common solar system issues 110

CHAPTER 11 113
ADVANCED SOLAR POWER TECHNIQUES 113
- Solar Power System Automation: Smart Home Integration 115
- Maximizing Efficiency with Solar Trackers 118
- Integrating Solar Power with Electric Vehicle Charging 120
- Advanced Troubleshooting and System Optimization 122

CHAPTER 12 125
FINANCIAL AND ENVIRONMENTAL IMPACT ANALYSIS 125
- Long-Term Financial Benefits of Solar Power 125
- Environmental Impact and Sustainability 127
- Case Studies: Successful Off-Grid Solar Installations 129
- Calculating ROI for Solar Investments 132
- Understanding Carbon Footprint Reduction 134

CHAPTER 13 137
SOLAR POWER FOR SPECIALIZED APPLICATIONS 137
- Solar Power for Tiny Homes and Portable Units 137
- Using Solar Power in Agricultural Settings 140
- Solar Water Pumping Systems 142
- Solar Energy for Emergency Preparedness 144
- Off-grid solar for Remote Locations 146

CHAPTER 14 151
NAVIGATING LEGAL AND REGULATORY CHALLENGES 151

- Understanding Solar Power Regulations 151
- Navigating Zoning Laws and Permits............................ 154
- Insurance Considerations for Solar Installations 156
- Leveraging Government Incentives and Subsidies....... 159
- Staying Compliant with Electrical Codes 161

CHAPTER 15 165
THE FUTURE OF SOLAR TECHNOLOGY 165

- Emerging Trends in Solar Technology 165
- Innovations in Solar Panel Materials 167
- The Role of Artificial Intelligence in Solar Power 170
- The Impact of Energy Storage Advancements 172

INTRODUCTION TO ENERGY INDEPENDENCE

In a world increasingly aware of the finite nature of its resources, the quest for energy independence has emerged as a beacon of hope and innovation. Energy independence is not just a concept; it's a transformative journey that empowers individuals, communities, and nations to seize control of their energy destinies. By harnessing renewable resources, we challenge the status quo and pave the way for a sustainable future, free from the constraints and volatility of traditional energy sources.

The significance of energy independence in the modern world cannot be overstated. This signifies a departure from the financial and ecological risks associated with fossil fuels, a move in the direction of lowering carbon emissions and strengthening the international effort to combat climate change. More than ever, the shift towards renewable energy sources, particularly solar power, is seen not just as an alternative but as an imperative for a healthier planet and a more secure energy future.

This chapter unfolds the rising popularity of solar power, reflecting a collective aspiration for a cleaner, self-sufficient, and resilient way of life. As we embark on this journey together, we invite you to envision a world where energy independence is not just a dream but a tangible reality, achieved through the power of the sun, innovation, and the indomitable human spirit. Let us illuminate the path to energy freedom, one solar panel at a time, unlocking the door to a sustainable legacy for future generations.

The Rising Popularity of Solar Power

The shift towards renewable energy has accelerated in recent years, with solar power shining as a leading light in this global transition. This growing trend is driven by a confluence of factors that underscore the urgency and viability of solar energy as a cornerstone for a sustainable future. The allure of solar power lies not only in its environmental benefits but also in its increasing economic viability, technological advancements, and the empowering promise of energy independence it offers to individuals worldwide.

The environmental imperatives of our time have catalyzed the surge in solar power's popularity. As the reality of climate change becomes more undeniable, solar energy stands out as a clean, inexhaustible source of power that significantly reduces greenhouse gas emissions. This alignment with global sustainability goals has propelled solar power from a niche alternative to a mainstream energy solution, heralding a new era of eco-conscious energy consumption.

Economically, the landscape of solar power has transformed dramatically. The price of solar panels and other related technologies has significantly dropped. Making solar installations more accessible to homeowners, businesses, and even nations. This cost reduction, coupled with the rising efficiency of solar panels, has made solar energy an increasingly competitive alternative to fossil fuels, promising long-term savings and a reduced dependency on fluctuating energy markets.

Technological advancements have further fueled the rise of solar power. Innovations in photovoltaic materials, energy storage solutions, and system integration have enhanced the efficiency and reliability of solar power systems, broadening their appeal and application. From remote off-grid locations to urban rooftops, solar technology is being adapted to meet diverse energy needs with unparalleled flexibility and scalability.

The promise of energy independence is the most compelling driver behind the rising popularity of solar power. Solar power offers individuals and communities the means to produce their own clean, sustainable energy in a world seeking resilience in the face of geopolitical tensions and energy security concerns. This autonomy reduces reliance on grid power and fossil fuels and empowers people to take control of their energy futures, fostering a sense of self-sufficiency and resilience.

Solar power emerges as a beacon of hope and progress as we stand on the cusp of a renewable energy revolution. Its rising popularity reflects a collective yearning for a cleaner, more sustainable, and more empowered world. This chapter sets the stage for a detailed exploration of solar power, inviting readers to join in the journey toward a brighter, solar-powered future.

CHAPTER 1

SOLAR POWER FUNDAMENTALS

Solar power is at the heart of the renewable energy movement, a technology that harnesses the sun's energy to generate clean, sustainable electricity. Anyone interested in investigating solar power's potential as a primary energy source should have a basic understanding of the technology. This section lays the foundation, covering the basic principles, components, and variations of solar power systems and providing a comprehensive overview for enthusiasts and future adopters alike.

Solar energy originates from the sun's radiation, a powerful and inexhaustible energy source that reaches the Earth daily. Photovoltaic (PV) cells, made from semiconductor materials, are the building blocks of solar power systems. These cells use the photovoltaic effect to convert sunlight directly into electricity, where light energy causes electrons in the semiconductor to generate an electric current.

A solar power system typically comprises several key components: solar panels, inverters, charge controllers, battery storage, and mounting structures. Solar panels and PV cell arrays are the fundamental components that collect sunlight and effectively convert it into direct current (DC) electricity. Inverters are vital in transforming this DC electricity into alternating current (AC), the standard form used in homes and businesses. Charge controllers manage how much power goes into and out of battery storage devices. Ensuring batteries are charged efficiently and protected from overcharging. Battery storage allows excess energy to be stored when sunlight is unavailable, providing a continuous power supply—finally, mounting structures secure solar panels in optimal positions to capture sunlight.

Solar power systems can be classified into two main types: off-grid and on-grid. Off-grid, or standalone systems, operate independently of the electricity grid, providing a self-sufficient energy solution ideal for remote locations. Grid-tied systems, also known as on-grid systems, are connected to the public power grid. The excess energy produced by these systems can be fed back into the grid, and the owner of the system may receive compensation for it.

The advantages of adopting solar power are manifold. Beyond its environmental benefits, such as reducing carbon footprint and conserving finite fossil fuels, solar energy can offer significant financial savings over time. With the cost of solar technologies continuing to decrease, solar power has become an increasingly accessible and cost-effective energy solution for many applications.

The Science of Solar Energy

It's a fascinating voyage into physics and engineering to understand how solar panels transform sunshine into power, but the process is explicable. At the core of this process is the photovoltaic (PV) effect, a principle that allows solar panels to convert light (photons) into electricity (voltage and current). Here's a breakdown of this process in an easy-to-understand manner.

Photovoltaic (PV) Effect

Absorption of Sunlight: Each solar panel is composed of many solar cells made from layers of semiconductor materials, typically silicon. Photons, or light energy, are absorbed by semiconductor material in solar cells when sunlight strikes them.

Generation of Electron-Hole Pairs: The energy absorbed from the photons excites electrons in the semiconductor material, giving them enough energy to break free from their atoms. This creates what are known as "free electrons" and leaves behind "holes." Free electrons and holes in the material create potential electrical energy.

Movement of Electrons: Solar cells are designed with a built-in electric field that causes these free electrons to move in a specific direction. This electric field is established by putting two layers of semiconductor material together, one with a positive charge (p-type) and one with a negative charge (n-type). The interaction between these layers creates the electric field at the junction.

Creation of Electrical Current: As the free electrons move along the electric field, they flow through the material to the contacts on the side of the cell. This flow of electrons is what we know as electricity. Connecting the external contacts to a load, such as a light bulb or a battery, the electrons flow through the circuit, powering the connected device before returning to the solar cell to fill in the holes, completing the circuit.

Conversion to Usable Power: The electricity generated by solar panels is direct current (DC). Most homes and appliances use alternating current (AC), so an inverter converts the DC into AC, making the electricity generated by solar panels usable for everyday needs.

Solar panels convert sunlight into electricity by absorbing photons, which excite electrons, creating an electrical current within a built-in electric field. Despite having intricate scientific and engineering roots, this method uses fundamental physics to provide a sustainable energy source that can run businesses, homes, and other structures. Through the marvel of the photovoltaic effect, solar energy offers a clean, sustainable solution to our energy needs.

Components of a Solar Power System

A solar power system is made up of multiple essential parts that work together to transform sunshine into electricity that can be used. Each part is crucial in ensuring the system operates efficiently and safely. Here's an overview of the main components and their functions:

Solar Panels

Solar panels, or photovoltaic (PV) panels, are the heart of a solar power system. They collect solar radiation and use it to generate DC electricity. Made from silicon cells, glass, a metal frame, and wiring layers, solar panels absorb photons from sunlight, initiating the photovoltaic effect that generates electric current.

Batteries (for off-grid and hybrid systems)

When solar panels are unable to produce enough electricity during the day, batteries store the energy produced by the panels for later use. This allows energy to be available at night or on cloudy days. They are essential for off-grid systems and can also be used in grid-tied systems with battery backup. Batteries allow for energy independence and ensure a continuous power supply.

Inverters

An inverter converts the DC electricity produced by solar panels into alternating (AC) electricity, the standard electrical current used by household appliances and the grid. Since solar panels have DC, and most home appliances operate on AC, inverters are critical for making the electricity generated by the solar panels usable for everyday needs.

Charge Controllers

The voltage and current that go from the solar panels to the batteries are controlled by charge controllers. They protect batteries from being overcharged in sunny conditions and prevent them from discharging too much in low-light conditions. Charge controllers are vital for battery longevity and efficiency, ensuring batteries are charged optimally and maintained properly.

Additional Components

Mounting Systems: Structures that secure solar panels in place on a roof or the ground. They are made to endure weather while maximizing the panels' tilt and position to receive as much sunlight as possible.

Wiring and Connectors: Electrical components that connect all parts of the solar power system, ensuring electricity's safe and efficient flow.

Electrical Disconnects, Breakers, and Fuses: Safety devices that protect the solar power system and home from electrical issues. They allow for safe maintenance and repair by disconnecting the solar power system from the grid and other electrical sources.

Monitoring Systems: Devices that track the performance of a solar power system, including energy production and consumption, system health, and efficiency. They can help identify issues early and optimize the system's performance.

Together, these components form a solar power system that can harness the sun's energy, convert it into usable electricity, and provide a renewable, sustainable power source for residential, commercial, or industrial applications.

Off-Grid vs. On-Grid

When contemplating the shift to solar power, one of the fundamental choices homeowners face is deciding between an off-grid and an on-grid solar power system. This decision significantly influences the installation's design, cost, and operation, tailoring it to specific energy needs and independence goals.

An on-grid, or grid-tied, solar power system is directly connected to the public electricity grid. This setup allows homeowners to feed excess electricity from their solar panels back into the grid, often receiving credits through net metering policies. The primary advantage of being grid-tied is using the grid as a virtual battery, eliminating the need for physical storage. This system is cost-effective and simplifies installation and maintenance. However, reliance on the grid means that in the event of a power outage, the solar power system will also shut down unless it's equipped with a specific type of inverter that allows for islanding or is part of a system designed to work during outages.

Conversely, off-grid solar power systems operate independently of the public electricity grid. This autonomy requires a comprehensive setup, including solar panels, battery storage, charge controllers, and often, additional generators for backup. Complete energy independence is the main attraction of off-grid living, which makes it perfect for isolated areas where grid connection is either unfeasible or too expensive. The major challenge of going off-grid is ensuring the system is adequately designed to meet all energy needs throughout the year. This can require significant upfront investment in components and careful planning to avoid power shortages.

Advantages of Off-Grid Living

Both systems have their merits and limitations. On-grid systems offer simplicity, cost savings, and the security of having the grid as a backup. In contrast, off-grid systems provide total independence, making them appealing to those seeking self-sufficiency or living in remote areas. The choice between on-grid and off-grid solar depends on individual priorities, such as the desire for independence, the availability of the grid, investment capacity, and specific energy needs. Making an informed decision involves weighing these factors against local regulations, available incentives, and the practicality of maintaining a solar power system in one's unique living situation, environmental footprint, and potential cost savings over time. This lifestyle is about energy independence and embodies a broader commitment to sustainability and self-reliance.

Autonomy and Self-Sufficiency

One of the most compelling benefits of off-grid living is its complete autonomy. Individuals and families become their power providers, free from reliance on utility companies and the vulnerability of grid outages. This independence means that fluctuations in energy prices or service disruptions do not impact those living off-grid. Instead, they enjoy a consistent power supply from their solar systems, tailored to meet their energy needs. This self-sufficiency extends beyond electricity, often including water sourcing and waste management, further reducing dependence on municipal services.

Reduced Environmental Impact

Living off-grid significantly reduces one's carbon footprint, a critical consideration in an era of environmental consciousness. Off-grid solutions prevent greenhouse gas emissions linked to fossil fuel-based electricity generation by using solar energy. This cleaner energy choice contributes to the global effort to combat climate change, preserve natural resources, and promote biodiversity. Furthermore, off-grid living encourages a lifestyle that values efficiency and sustainability, often incorporating eco-friendly practices like composting, rainwater harvesting, and sustainable agriculture.

Potential Cost Savings

While the initial setup of an off-grid solar system can be substantial, the long-term cost savings can be significant. Eliminating monthly electricity bills and reducing other utility costs means that, over time, the investment in solar panels, batteries, and other components pays off. Additionally, off-grid homeowners are not subject to the increasing utility electricity rates or the costs associated with infrastructure maintenance and upgrades. The financial benefits extend beyond mere savings, as living off-grid can also increase self-awareness of energy consumption, leading to more conscientious use and further financial and environmental benefits.

Embracing a Sustainable Lifestyle

Beyond the tangible benefits, off-grid living often reflects a deeper commitment to a sustainable and mindful lifestyle. It fosters a closer connection to the natural environment, encouraging practices that respect and preserve our planet's resources. This lifestyle choice can offer a profound sense of fulfillment, knowing that one's living habits contribute positively to the environment and future generations.

How You Can Share Your Review

Through Amazon.com

- Visit the Amazon page where you purchased or found my book.
- Scroll down to the 'Customer Reviews' section near the bottom of the page.
- Click on 'Write a customer review' to begin sharing your valuable insights and experiences with the book.

Instant QR Code Access

Simply scan the QR code below with your smartphone to be directly taken to the Amazon review section for the book. This quick access method makes it easy for you to leave your feedback without navigating through the website.

Your review is not just feedback for us; it's a beacon for future readers navigating the world of solar energy and sustainable living. Thank you for taking the time to share your thoughts and helping us spread the message of sustainability.

CHAPTER 2

INITIAL CONSIDERATIONS FOR GOING SOLAR

Assessing Your Energy Needs

Assessing your energy needs is a critical first step in transitioning to solar power, ensuring that your system is neither underpowered nor excessively large for your actual requirements. This involves understanding your current energy consumption patterns to design a solar power system that aligns with your lifestyle or business needs. Here's a straightforward guide to help you calculate your energy consumption and pave the way for a system that meets your needs efficiently.

Gather Your Utility Bills

Start by collecting your electricity bills for the past year. These documents provide a comprehensive overview of your energy consumption over different seasons, accounting for variations in usage due to heating, cooling, and other seasonal demands. Look for the kilowatt-hours (kWh) you've consumed each month, as this unit of measurement is key to understanding and calculating your energy needs.

Calculate Your Average Energy Use

Once you have your yearly consumption data, calculate your average monthly and daily energy use. This can be done by adding up the total kWh used over the year and dividing by 12 for the monthly average or 365 for the daily average. This step provides a baseline for understanding how much energy your solar power system needs to generate to meet your demands.

Factor in Efficiency and Future Needs

Considering the efficiency of appliances and any changes in your future energy consumption is essential. If you plan to switch to more energy-efficient appliances or add new loads, such as electric vehicle charging, these factors should be incorporated into your calculations. Additionally, consider any planned changes to your lifestyle or property that might increase your energy needs.

Use Online Calculators for Precision

Numerous online solar energy calculators can help refine your energy needs assessment. These tools often factor in local solar irradiance data, system losses, and the efficiency of different solar panel types, offering a more detailed and customized analysis of your specific situation.

Professional Energy Audit

For a more comprehensive assessment, consider a professional energy audit. An auditor can identify inefficiencies in your home or business and recommend reducing energy consumption before installing a solar system. This can lead to a more minor, cost-effective system by lowering overall energy needs.

Project Your Solar System Size

It is possible to determine the approximate size of the solar power system needed to suit your demands by using the average daily energy usage statistic. This entails taking into account the efficiency of the solar panels you intend to employ as well as the typical number of peak sunshine hours that your area receives. The formula for a fundamental estimation is:

System Size (kW) = Daily Energy Consumption (kWh) Average Peak Sunlight Hours per Day Inverter Efficiency Factor

Keep in mind that this calculation provides an estimate. Other factors, such as the angle and orientation of the solar panels, potential shading, and the type of solar technology used, will also influence the final system size recommendation.

By carefully assessing your energy needs, you lay the groundwork for a solar power system that is tailor-made for your specific situation, ensuring optimal performance and financial savings over the lifespan of your solar installation.

Solar Potential and Climate Considerations

A location's capacity for producing solar electricity is greatly influenced by its topography. This encompasses physical geography, such as latitude and terrain, and climate factors, including the amount of sunlight, cloud cover, and temperature variations. Understanding how these elements affect solar power generation is essential for maximizing the output of a solar energy system.

Latitude and Sunlight Intensity

The mean solar radiation intensity changes with latitude due to the Earth's tilt and orbit around the sun. The best places to generate solar electricity are those that are closest to the equator since they receive more direct sunshine there all year round. However, even higher latitude locations can harness solar energy effectively, especially with technologies designed to capture diffuse light on overcast days.

Terrain and Solar Access

The terrain of a location can significantly impact solar power generation. Areas with open access to the sky without obstruction from mountains, buildings, or trees receive more sunlight, enhancing solar panel efficiency. Conversely, locations in valleys or areas prone to shading may require strategic placement of solar panels or additional capacity to meet energy needs.

Climate and Weather Patterns

Climate plays a pivotal role in solar power generation. Regions with high levels of sunshine and clear skies are naturally more suited to solar energy production. However, even areas with frequent cloud cover can benefit from solar power, thanks to photovoltaic (PV) technology advancements that improve efficiency in various weather conditions. It's also worth noting that solar panels operate more efficiently in cooler temperatures, making some less sunny, cooler regions surprisingly effective for solar energy.

Maximizing Solar Power Output

To maximize solar power output, consider the following strategies tailored to your geographic and climatic context:

Optimal Panel Orientation and Tilt: Aligning solar panels to face the true south (in the northern hemisphere) and adjusting their tilt angle according to your latitude can significantly increase energy production.

Seasonal Adjustments: For locations with significant seasonal variations, adjusting the tilt of the panels at different times of the year can capture more sunlight.

Technology Choice: Using PV panels that perform well in your specific climate, such as those optimized for low light in cloudier climates or with high-temperature coefficients for hot areas, can enhance efficiency.

Microclimate Considerations: Pay attention to local microclimates created by landscape features and built environments, and position solar panels to avoid shaded areas.

Monitoring and Maintenance: Regularly monitoring system performance and maintaining panels clean and free from obstructions like snow or debris ensures optimal efficiency.

You can significantly improve their efficiency and output by carefully considering geography and climate in the design and placement of solar power systems. Tailoring the system to your specific location maximizes energy production and increases the overall return on investment in solar technology.

Budgeting for Your Solar Project

Understanding and planning the financial aspect of your solar project is crucial for a smooth and successful transition to solar energy. This includes the upfront cost of solar panels and installation and potential hidden costs that can arise during or after the project. By carefully budgeting for these expenses, you can ensure a more accurate financial plan and avoid surprises.

Start by getting quotes from multiple solar installation companies to compare prices and services. This will give you a general idea of the upfront costs for your solar system. Remember, the cheapest option is only sometimes the best. Consider the quality of the components, the reputation of the installer, and the warranties offered.

Factor in potential hidden costs such as permits, taxes, and fees. Depending on your location, you may need to pay for building permits, inspections, and other regulatory requirements. Additionally, some utilities charge fees for connecting your solar system to the grid.

Consider the cost of financing if you need to pay for your solar system outright. Loans and solar leases can add interest and fees to the total cost of your system, so it's important to compare different financing options and understand the long-term financial implications.

Remember maintenance and repair costs. While solar panels require minimal maintenance, you should budget for occasional cleaning, potential repairs, and replacing components like inverters or batteries over the lifespan of your system.

Finally, research available incentives and rebates. Many governments offer tax credits, rebates, and other incentives to reduce the cost of going solar. These can greatly reduce the initial outlay of funds, but be aware of the qualifications and procedures for claiming them.

It is possible to develop a more precise and practical financial strategy for your solar project by carefully accounting for both the obvious and unforeseeable expenditures. This helps manage your finances effectively and maximizes the return on your investment in solar energy.

Understanding Your Site's Solar Access

In order to optimize the energy production of your solar system, it is imperative that you evaluate your property for the best location for solar panels. The quantity of sunshine that a solar power system can use to generate energy is known as solar access, and it has a big influence on how successful and efficient your solar power setup is. Here's how to evaluate the solar access at your location:

Observe Sunlight Patterns

Start by observing the sunlight patterns over your property throughout the day.

Identify areas that receive consistent, unobstructed sunlight, especially during peak sunlight hours (generally between 9 AM and 3 PM).

Note any obstacles like buildings, trees, or landscape features that might cast shadows during these hours.

Consider Roof Orientation and Angle: The effectiveness of your solar panels is greatly influenced by the direction and angle of your roof. South-facing roofs are the best in the Northern Hemisphere since they get the most sunshine all day long. The angle of your roof should also be considered; solar panels perform best when perpendicular to the sun's rays. An angle that matches your location's latitude is a good starting point, though adjustments may be necessary based on specific site conditions.

Use Solar Mapping Tools: Numerous online tools and software can help assess your property's solar potential. These tools analyze satellite imagery and weather data to estimate how much sunlight your roof receives and can even account for shading from nearby obstacles. They can provide a rough estimate of potential solar energy generation, helping to identify the most promising locations for solar panel installation.

Professional Site Assessment: For a more accurate analysis, consider hiring a professional solar installer to conduct a site assessment. They can perform a detailed shading analysis and evaluate factors affecting installation, such as roof condition, structural integrity, and local regulations. This assessment will provide tailored recommendations for maximizing solar access and system performance.

Understand Seasonal Variations: Remember that the sun's path changes with the seasons, affecting solar access throughout the year. Obstacles that don't cast shadows in summer may block sunlight in winter when the sun is lower in the sky. A comprehensive evaluation will consider these seasonal variations to ensure consistent solar production year-round.

By thoroughly understanding and evaluating your site's solar access, you can make informed decisions about the placement and design of your solar power system. This optimizes the efficiency of your solar panels and enhances the long-term benefits of your investment in renewable energy.

CHAPTER 3

NAVIGATING REGULATIONS AND INCENTIVES

Local Zoning and Permitting Process

Navigating the local zoning and permitting process is crucial in the solar installation journey, ensuring your project complies with all legal requirements and local regulations. While specifics can vary widely based on jurisdiction, here's a general overview to guide you through understanding and managing these requirements for your solar project.

Understanding Zoning Laws

Zoning laws dictate how land within certain areas can be used, impacting where and how solar panels may be installed on your property. These laws may specify restrictions related to the height of structures, setback distances from property lines, and even the aesthetic considerations of solar installations in historic districts or certain community developments.

Navigating the Permitting Process

The permitting process for solar installations typically involves submitting detailed plans of your proposed system to the local government or building department. These plans must often include specifications of the solar panels, wiring diagrams, mounting methods, and an outline of how the system integrates with the existing electrical setup. The goal is to ensure safety, structural integrity, and compliance with electrical codes.

How to Find Specific Information for Your Area

Contact Your Local Building Department or Government Office: They can provide detailed information on your area's zoning laws and permit requirements. They may also offer solar installation guidelines, checklists, and application forms.

Consult a Professional Solar Installer: Solar installers with experience in your area can be invaluable resources. They are usually familiar with the local zoning laws and permitting process and often handle permit applications as part of their services.

Review Homeowner Association (HOA) Rules: If your property is included in an HOA, review the rules and regulations of the association. There are HOAs that have rules or procedures that must be followed before installing solar panels.

Check State and Federal Resources: Some states have online resources or helplines for solar energy projects. These can provide information on state-specific regulations, incentives, and tips for navigating local requirements.

Attend Local Workshops or Seminars: Local governments or environmental organizations occasionally host workshops or seminars on going solar. These events can offer helpful information and direct access to local officials or experts who can answer specific questions.

Common Requirements and Considerations

Building Permits: Most jurisdictions require a building permit to install a solar panel system. The application process ensures your project meets local building codes and safety standards.

Electrical Permits: An electrical permit is often required because solar installations involve electrical work. This ensures the electrical components of your solar system are up to code.

Inspections: After installation, your solar system will likely need to pass one or more inspections to verify compliance with local regulations and permit requirements. Inspectors will check the electrical wiring, mounting system, and overall installation integrity.

By familiarizing yourself with the local zoning and permitting process and leveraging available resources, you can ensure your solar installation proceeds smoothly and legally. While the process may seem daunting initially, proper planning and consultation with professionals can significantly simplify compliance with local regulations, paving the way for a successful solar project.

Federal and State Solar Incentives

Knowing the available federal and state solar subsidies might help you lower the installation costs substantially when you make the switch to solar electricity. By lowering the cost of solar electricity for both homes and businesses, these incentives aim to promote the use of renewable energy.

The Investment Tax Credit (ITC), which lets you deduct a certain portion of the cost of your solar system from your federal taxes, is one of the most beneficial government incentives. This substantial savings can dramatically lower the initial investment required for solar installation. The specifics of the ITC, including the percentage of cost that can be claimed, may vary over time, so it's crucial to check the current details to understand how it can benefit your project.

Rebates, tax breaks, and certifications of renewable energy from sunlight (SRECs) are just a few examples of the many different state incentives that are available. State tax credits work similarly to the federal ITC, providing a deduction from your state taxes. Rebates, offered by some states, utilities, or local governments, directly reduce the upfront cost of your solar system. SRECs allow you to earn extra income by selling certificates associated with the amount of electricity your system generates.

To explore these prospects, the Database of State Incentives for Renewables & Efficiency (DSIRE) is a useful place to start. This extensive website offers full details on the regulations and incentives that encourage renewable energy in the US. You may obtain a customized list of local solar incentives, along with application criteria and thorough explanations, by entering your zip code.

Additionally, consulting with a professional solar installer can provide clarity and guidance. Many installers are well-versed in the nuances of local, state, and federal incentives and can assist you in maximizing the financial benefits for your specific situation.

By taking advantage of these incentives, you can significantly offset the cost of going solar, making it a more accessible and appealing investment. It's worth investing the time to research and understand the incentives available to you, as they can play a crucial role in your solar power project's overall affordability and success.

How to Claim Solar Tax Credits

Claiming solar tax credits can significantly reduce the overall cost of your solar power system, making renewable energy more accessible and affordable. There are a few phases in the procedure, but if you plan ahead and pay close attention to details, you may complete the claim process without any problems. Here's a step-by-step guide to help you claim available solar tax credits:

Step 1 Verify Eligibility

First, ensure your solar power system qualifies for the solar tax credit. Generally, the system must be installed at your primary or secondary residence in the United States and be used to generate electricity for the residence. The tax credit can be claimed for new installations and certain costs associated with solar system installations.

Step 2 Gather Documentation

Collect all necessary documentation related to the purchase and installation of your solar power system. Invoices, receipts, and evidence of payment (such as credit card or bank statements) are included in this that clearly show the cost of the solar panels, installation charges, and any additional eligible expenses. Also, keep a copy of the Manufacturer's Certification Statement for your records, if applicable.

Step 3 Determine the Credit Amount

Calculate the amount of tax credit you are eligible to claim. By using the federal solar tax credit, you may deduct a portion of the solar energy system installation costs from your federal taxes. The percentage can change, so refer to the current tax year's guidelines to determine the applicable rate.

Step 4 Complete IRS Form 5695

For federal tax credits, complete IRS Form 5695, "Residential Energy Credits." This form calculates the credit you can claim for your solar power system. Fill in the details of your expenses in the appropriate sections to determine the credit amount. If you also claim credits for other energy improvements, ensure they are correctly accounted for on the form.

Step 5 Attach to Your Tax Return

Once you have completed Form 5695, include the credit amount on your 1040 form. Attach Form 5695 to your tax return when you file it. If you're using tax software, the program should guide you through entering your solar tax credit information and attaching the necessary forms electronically.

Step 6 Carry Over Unused Credits

You can carry over the excess credit to the following tax year if it exceeds your annual tax liability. Check the current tax laws to understand how carryover works and ensure you're taking full advantage of the credit available.

Step 7 State and Local Credits

Additionally, investigate any state or local tax credits you may be eligible for. The process for claiming these can vary by jurisdiction, so consult your state or local tax authority or a tax professional familiar with renewable energy credits in your area.

Step 8 Seek Professional Advice

To assist in navigating the complexity of tax incentives for solar systems, think about speaking with a tax expert. They may offer tailored guidance according to your circumstances, making sure you get the most credit possible.

These procedures will allow you to successfully claim solar tax credits, lower the cost of your solar installation, and make sustainable energy more financially viable.

Net Metering and Other Incentives

Net metering and other incentives significantly enhance the attractiveness and viability of solar power systems for homeowners and businesses. These incentives reduce the upfront costs of going solar and provide ongoing benefits that can make solar energy a financially rewarding investment over time.

Net Metering

Owners of solar panels benefit greatly from net metering, which enables them to return surplus energy produced by their panels to the grid in exchange for credits. This is how solar panel owners gain from it:

Reduced Utility Bills: When the energy generated by your solar panels exceeds your use, the excess is exported to the grid and credited to your account. When your solar system isn't producing enough power to fulfill your needs, such as at night or on cloudy days, these credits can then be used to offset the cost of electricity received from the grid.

Optimized Investment: By effectively using the grid as a battery, homeowners can avoid the high costs and maintenance associated with physical battery storage systems. This maximizes the use of solar energy produced and ensures that the investment in solar panels pays off more quickly.

Contribution to Grid Stability: Net metering also benefits the wider community and the utility company by providing additional electricity during peak production times, which can help stabilize the grid and reduce reliance on fossil fuels.

State and Local Incentives

Beyond net metering, various state and local incentives further enhance the appeal of solar energy.

- **Rebates and Grants:** Some states, municipalities, or utility companies offer rebates or grants that directly lower the solar power system's initial installation costs. These incentives can significantly lower initial investment costs and improve the return on investment.

- **Tax Credits and Exemptions:** In addition to federal tax credits, some states offer tax incentives for solar energy. These may include state tax credits that directly reduce your state tax liability or property tax exemptions that prevent your property taxes from increasing due to adding a solar power system.

- **Solar Renewable Energy Certificates (SRECs):** In certain states, solar panel owners can earn SRECs for the electricity their system generates. These certificates can then be sold on a market to utilities needing to meet renewable energy requirements, providing an additional income stream for solar panel owners.

Feed-in Tariffs

Some regions offer feed-in tariffs (FITs), which pay solar panel owners a fixed rate for the electricity they generate and feed into the grid over a set period. This rate is usually higher than the retail or wholesale price of electricity, providing a steady income and a strong incentive for investing in solar energy.

Performance-Based Incentives (PBIs)

PBIs pay solar panel owners based on the actual electricity their systems produce over time. Unlike FITs, which pay a fixed rate, PBIs may vary in payment structure and rate but offer another way to monetize the energy produced by your solar panels.

By taking advantage of net metering and other available incentives, solar panel owners can significantly reduce their energy costs, accelerate the payback period of their solar investment, and even generate ongoing income. These incentives not only make solar power more accessible but also encourage the adoption of renewable energy, contributing to environmental sustainability and energy independence.

CHAPTER 4

STEP 1 DESIGNING YOUR SYSTEM

Determining System Size and Capacity

Determining the appropriate size and capacity of your solar power system is crucial in ensuring that your solar installation meets your energy needs efficiently and effectively. This process involves calculating your current energy consumption, anticipating future changes in usage, and considering your site's specific conditions to design a manageable and manageable system.

Start by examining your electricity bills over the past year to get a clear picture of your total energy usage. This gives you a baseline of how much power you need to generate to cover your consumption. Remember, the goal is to match your annual energy usage with the production of your solar system, keeping in mind any expected increases in consumption due to lifestyle changes or the addition of new appliances or electric vehicles.

Next, assess the solar potential of your location. Factors such as geographic location, the amount of daily sunlight your property receives, and seasonal variations significantly affect how much energy your solar panels can produce. You can get an idea of the possible energy output by estimating the solar irradiation in your location with the use of online tools and calculators.

Consider also the physical constraints of your installation site. The available space for solar panels, whether on your roof or the ground, will limit the size of your system. The orientation and tilt of the panels, as well as potential shading from trees, buildings, or other obstructions, will affect your system's efficiency and, consequently, its required size to meet your energy needs.

With this information, you can begin to calculate the size of the solar system required. The calculation involves dividing your total annual electricity usage by the expected annual energy production per unit of solar panel capacity, adjusted for inefficiencies and losses. This will give you an estimate of the total capacity needed to meet your energy demands.

Finally, consulting with a professional solar installer who can provide a detailed assessment and help refine your system size based on technical evaluations and experience is advisable. They can consider factors that might be obscure, such as local regulations, the impact of future shading as trees grow, or the potential for system expansion.

Determining your solar power system's right size and capacity is about more than just meeting your current energy needs. It's also about planning for the future and ensuring that your investment is as effective as possible over the lifetime of the solar installation. By carefully assessing your energy usage, solar potential, and site conditions, you can design a solar power system that offers maximum efficiency, sustainability, and cost savings.

Selecting the Right Solar Panels

Designing a solar power system that aligns with your energy needs, balances efficiency with cost, and integrates seamlessly into your property requires a systematic approach. From determining the appropriate system size and capacity to selecting the right solar panels, each step is critical to achieving a solar setup that meets your expectations for performance and budget.

Determining System Size and Capacity

The journey to a tailored solar power system begins with understanding your energy consumption. Analyzing your electricity bills over the past year will give you a baseline of how much energy you use, allowing you to calculate the solar system size needed to satisfy your demands. Consider the fluctuations in energy usage across seasons to ensure your system is adequately sized to handle peak demands.

Once you have a clear picture of your energy requirements, factor in the solar potential of your location. The amount of sunlight your property receives is influenced by its location, orientation, and any shading from nearby obstacles. This information, combined with data on the average sunlight hours available in your region, will help refine your system's capacity to generate sufficient power throughout the year.

Selecting the Right Solar Panels

With a grasp on the system size and capacity needed, the next step is selecting appropriate solar panels. There are several varieties of solar panels, and each has a unique set of efficiency, durability, and cost characteristics. Monocrystalline solar panels are renowned for their excellent efficiency and stylish design. It is an excellent choice for those with limited roof space, albeit at a higher price point. Polycrystalline panels balance performance and affordability, making them a popular option for residential installations.

When selecting solar panels, consider not only their initial cost but also their performance over time. Although they may cost more upfront, high-efficiency solar panels have the potential to produce more electricity over time, which might result in larger long-term cost savings on your energy bills. Additionally, the warranty and lifespan of the panels are crucial factors. A more extended warranty period and a robust build can protect your investment for years to come.

Balancing efficiency and cost involves carefully evaluating your immediate budget against the long-term benefits of your solar power system. While it may be tempting to opt for the least expensive option initially, investing in higher-quality components can lead to better performance, fewer maintenance issues, and more significant savings over the lifespan of your solar installation.

Designing your solar power system is a nuanced process beyond merely choosing equipment. It requires an understanding of your energy needs, the solar potential of your property, and the market's available technology. By thoughtfully considering each of these aspects, you can craft a solar power system that meets your current energy demands and adapts to future changes, ensuring a sustainable and cost-effective solution for years to come.

Charge Controllers PWM vs. MPPT

In solar power systems, selecting a suitable charge controller is crucial for managing the energy flow between your solar panels and the battery bank. Charge controllers protect batteries from overcharging and excessive discharge, extending their lifespan and ensuring the efficiency of your solar setup. Two primary types of charge controllers dominate the market: Pulse Width Modulation (PWM) and Maximum Power Point Tracking (MPPT). Understanding the differences between these two can help you make an informed decision that balances efficiency and cost.

Pulse Width Modulation (PWM) Charge Controllers

PWM charge controllers are the more traditional form of charge regulation. They work by slowly reducing the amount of power going into your battery as it nears capacity, employing a series of short charging pulses. The width of these pulses changes based on the battery's charge level, effectively reducing the charge current to prevent overcharging.

PWM controllers are simpler in design and generally less expensive than their MPPT counterparts. They are most effective when the solar panel's voltage closely matches the battery voltage, making them a suitable choice for smaller systems where the budget is a primary concern and system efficiency is not maximized.

Maximum Power Point Tracking (MPPT) Charge Controllers

MPPT charge controllers represent a more advanced technology designed to maximize the efficiency of the solar power system. These controllers continuously adjust the modules' or array's electrical operating point to ensure it delivers the maximum power available. MPPT controllers are particularly effective in varied weather conditions, including partial shading, low light, and cold temperatures, where they can significantly improve the energy harvest. Although MPPT controllers come at a higher initial cost, their ability to optimize power extraction from the solar panels can result in higher system efficiency and quicker return on investment, especially in larger or more complex solar installations.

Choosing Between PWM and MPPT

The choice between PWM and MPPT charge controllers should be guided by several factors, including the size of your solar power system, the conditions under which it will operate, and your budget constraints. A PWM controller may suffice for small, simple systems with a tight budget, offering a cost-effective solution without significantly compromising performance. On the other hand, for larger systems, those in areas with variable weather conditions, or when the panel voltage is considerably higher than the battery voltage, an MPPT controller can enhance performance and energy yield, making it a worthwhile investment despite its higher upfront cost.

Ultimately, the decision between PWM and MPPT charge controllers involves weighing the initial investment against the potential for increased energy efficiency and system performance. By carefully considering your specific needs and the characteristics of your solar installation, you can select a charge controller that ensures the optimal balance of efficiency, cost, and reliability, paving the way for a successful solar energy project.

Battery Storage Options

When integrating a solar power system into your home or business, considering your battery storage options is essential for maximizing energy independence and efficiency. Battery storage allows you to capture and store excess electricity produced by your solar panels for use when the sun isn't shining, such as during the evening or on cloudy days. This capability not only enhances the utility of your solar system but can also provide backup power during outages, making it a critical component for those seeking full energy autonomy.

Several types of battery storage technologies are available, each with its advantages and considerations. Lead-acid batteries, used for decades in off-grid solar systems, are known for their reliability and lower upfront cost. However, they require regular maintenance and have a shorter lifespan compared to other technologies, which can lead to higher costs over time.

Lithium-ion batteries have become increasingly popular for solar energy storage due to their higher efficiency, longer lifespan, and compact size. While the initial investment in lithium-ion technology is higher than in lead acid, the longer lifecycle, greater depth of discharge, and minimal maintenance requirements often result in a lower total cost of ownership. Additionally, lithium-ion batteries are capable of handling higher power loads, making them suitable for both residential and commercial applications.

Another option to consider is flow batteries, which store electricity in liquid electrolyte solutions. Though less common in residential solar installations, flow batteries offer several benefits, including a long lifespan, scalability, and the ability to discharge 100% of their stored energy without damage. However, they are typically more extensive and complex systems, requiring more space and a higher initial investment.

When selecting a battery storage option for your solar power system, it's essential to consider the cost and how the system's characteristics align with your energy needs. Factors such as the capacity and power rating of the battery, its depth of discharge (the percentage of the battery's energy that can be used without harming its lifespan), and its round-trip efficiency (the percentage of energy that can be used of the amount of energy it took to store it) will all influence the system's overall performance and suitability for your application.

Moreover, integrating battery storage into your solar power system should be planned with an eye toward future needs. As your energy consumption grows or changes, having a scalable storage solution can provide the flexibility to adapt without requiring a complete system overhaul.

Ultimately, your solar system's choice of battery storage balances immediate costs, long-term benefits, and how well the system matches your specific energy usage patterns, financial objectives, and sustainability goals. With the proper storage solution, you can significantly enhance your solar energy system's reliability, efficiency, and independence, making it a powerful tool in your energy management arsenal.

Choosing an Inverter for Your Needs

Choosing a suitable inverter is a critical decision in the design of your solar power system. The inverter is a pivotal component that converts your solar panels' direct current (DC) electricity into alternating current (AC) electricity, the standard used by household appliances and the grid. There are several types of inverters, each with its features, benefits, and considerations. Making an informed choice requires understanding these differences and how they align with your energy needs.

Types of Inverters

String Inverters are the most common type in home solar power systems. String inverters are connected to a series (or "string") of solar panels. They are efficient, cost-effective, and suitable for systems where solar panels are installed in a single plane with minimal shading issues. However, if one panel in the string becomes shaded or fails, the performance of the entire string can be compromised.

Microinverters: Unlike string inverters, microinverters are installed at each solar panel. This setup allows each panel to operate independently, optimizing the energy production of each panel and minimizing the impact of shading or panel failure on the overall system performance. Microinverters are ideal for roofs with multiple angles or shading issues but are more expensive than string inverters.

Hybrid Inverters: Hybrid inverters combine the functionality of a standard grid-tied inverter with the ability to connect to a battery storage system. This type of inverter allows energy storage of excess power to be used later or during grid outages, offering increased energy independence. Hybrid inverters are an excellent choice for systems designed with energy storage in mind, but they are more expensive **than string inverters.**

Considerations for Choosing an Inverter

System Size and Layout: The size of your solar power system and the layout of your solar panels play a significant role in determining the most suitable inverter. String inverters may offer a cost-effective solution for larger, more uniform arrays. Microinverters can provide better performance and flexibility for complex roof layouts or shaded environments.

Future Expansion: If you anticipate expanding your solar power system, consider how your choice of inverter might accommodate this growth. Microinverters make it easier to add panels without redesigning the entire system.

Budget: Budget constraints are often a deciding factor. In comparison, microinverters offer performance advantages but come at a higher initial cost. Evaluate the long-term benefits of each option against your current budget to make a decision that balances cost with performance.

Energy Storage: If integrating energy storage or planning for a battery backup is essential to you, a hybrid inverter may be the best choice. This option provides the versatility to manage energy consumption and storage efficiently.

Warranty and Reliability: Finally, consider the warranty and reliability of the inverter. A longer warranty period can provide peace of mind and protect your investment over time. Research the manufacturer's reputation and read reviews to ensure you're choosing a product known for durability and reliable performance.

Selecting the right inverter for your solar power system is about balancing efficiency, cost, and your specific energy needs. By carefully considering the type of inverter that best suits your situation, you can optimize the performance of your solar installation, ensuring it delivers maximum benefits for years to come.

Balancing System Efficiency and Cost

Balancing system efficiency and cost is critical when designing a solar power system. Achieving this balance involves making informed choices about the components of the system and understanding how each decision impacts both the initial investment and the long-term savings. The goal is to create a system that fits within your budget, maximizes energy production, and efficiently meets your energy needs over time.

When considering the efficiency of a solar power system, the focus often shifts to the solar panels themselves. High-efficiency panels can produce more electricity from the same sunlight as standard panels. While they come with a higher price tag, the increased energy output can lead to greater savings on your energy bills, potentially offsetting the initial cost difference over the system's life. Therefore, investing in higher-efficiency panels might be wise if your roof space is limited or you want to maximize your energy production.

However, efficiency extends beyond just the panels. The choice of inverter, for example, plays a significant role in the system's overall efficiency. Inverters convert the DC electricity generated by the panels into AC electricity that can be used in your home or fed back into the grid. Selecting an inverter that matches the output of your solar panels and minimizes energy loss during the conversion process is crucial for maintaining system efficiency.

Battery storage options are also factored into the equation. Batteries allow you to store excess energy when solar production is low, enhancing the system's utility and independence. However, the efficiency of energy storage and retrieval, as well as the cost and lifespan of the batteries, need to be considered. More efficient batteries may cost more upfront but can provide better long-term value by reducing energy waste and extending the period between replacements.

In balancing efficiency and cost, it's also important to consider the potential for future expansion. Designing a system that can easily accommodate additional panels or batteries can save significant costs and disruption compared to overhauling the system later. This foresight allows for gradual investment in your solar power system, spreading costs while increasing energy independence and efficiency.

Ultimately, the key to balancing system efficiency and cost lies in a comprehensive evaluation of your energy needs, budget constraints, and the specific characteristics of your property. Consulting with solar energy professionals can provide valuable insights into the latest technologies and products that offer the best balance for your situation. Additionally, taking advantage of available incentives and rebates can further offset the initial costs, making a more efficient system accessible within your budget.

By carefully considering each component and making informed decisions, you can design a solar power system that meets your current energy needs and provides a scalable, efficient solution that will serve you well into the future. This balance between efficiency and cost ensures that your investment in solar power is both financially and environmentally sustainable.

CHAPTER 5

STEP 2 PROCURING COMPONENTS

The second step in setting up your solar power system involves procuring the right components. This phase is crucial as each element's quality, compatibility, and efficiency directly impact the overall performance and reliability of your solar setup. Navigating through the procurement process requires a strategic approach to select the best products within your budget, ensuring that your system meets both your immediate and long-term energy needs.

Begin by creating a comprehensive list of components needed for your solar power system, typically including solar panels, an inverter, batteries for storage, charge controllers, mounting hardware, and wiring. Each component comes in various types and specifications, making understanding their functions and how they fit into your system's design is essential.

When selecting solar panels, consider efficiency, durability, warranty, and the manufacturer's reputation. Panels with higher efficiency might cost more upfront but can generate more electricity over the same surface area, making them a wise choice for maximizing output, especially in limited spaces.

Choosing an inverter requires matching its capacity to your system's size and understanding the differences between string, microinverters, and hybrid inverters. The right choice depends on your system's configuration, the potential for expansion, and whether you plan to include battery storage.

For battery storage, evaluate the types of batteries available, such as lead-acid or lithium-ion, weighing their life expectancy, maintenance requirements, and cost-effectiveness. Your decision should align with your energy storage needs, considering how often you'll rely on stored power.

The procurement process extends to smaller, significant components like mounting systems and wiring. These should be chosen based on durability and compatibility with your installation environment to ensure a secure and long-lasting setup.

As you navigate procuring components, sourcing from reputable suppliers and manufacturers is vital. Look for vendors with positive reviews and robust warranties that offer protection against defects and performance issues. It's also worth comparing prices and seeking packages or discounts to purchase multiple components together.

Consider leveraging professional advice during this phase. Solar energy professionals can offer insights into the best component combinations for your situation and may provide access to higher-quality products at competitive prices. They can also help you navigate warranties and technical specifications, smoothing the procurement process.

Finally, keep an eye on available incentives and rebates for solar installations in your area. These can significantly reduce the overall cost of your components, making higher-quality options more accessible within your budget.

Procuring components for your solar power system is a delicate balance between quality, cost, and performance. By carefully selecting each element with an eye towards the future, you not only ensure the efficiency and reliability of your system but also protect your investment, ensuring that your solar setup meets your energy needs for years to come.

Where to Buy Solar Equipment

Deciding where to buy solar equipment is as crucial as selecting the components themselves, as it influences not just the cost but also the quality and reliability of your solar power system. There are several avenues through which you can acquire solar panels, inverters, batteries, and other necessary equipment, each with its own set of advantages and considerations.

Specialized vendors of solar energy equipment are one of the main places to buy solar equipment. These sellers frequently offer a large selection of goods, making it possible to compare various brands and technological advancements. They typically offer expert advice based on your specific project requirements, which can be invaluable for those new to solar power. Additionally, purchasing from a reputable supplier ensures that the equipment comes with manufacturer warranties and support, providing peace of mind and protection for your investment.

Another option is to buy directly from the manufacturers. Some solar equipment manufacturers sell now to consumers, often through their websites. Buying now can sometimes result in cost savings by cutting out the middleman. Moreover, it offers the opportunity to receive detailed product information and support directly from the people who made the equipment. However, direct purchases may require a deeper understanding of the solar components to ensure compatibility and meet your energy needs effectively.

Online marketplaces and retailers represent a third option, offering the convenience of shopping from home and the ability to easily compare prices and reviews. While often competitive in terms of pricing, it's essential to exercise caution to ensure you're buying genuine, warranty-backed products. Pay close attention to seller reviews, product warranties, and return policies to avoid counterfeit or substandard equipment.

Local solar installers can also be a source for purchasing solar equipment. Many installers offer full-service packages that include equipment procurement based on a customized design for your home or business. This approach ensures that all components are compatible and optimized for your specific installation. Additionally, installers can handle the entire process from design and procurement to installation and maintenance, offering a hassle-free solution for those looking for a turnkey approach.

When deciding where to buy your solar equipment, consider factors such as cost, convenience, product range, and after-sales support. It's also wise to research and compare different vendors, taking note of customer experiences and feedback. Regardless of where you choose to make your purchase, ensuring the quality and reliability of the equipment should be paramount. Opting for products from established, reputable brands with solid warranties can safeguard your solar investment over the long term.

Evaluating Component Quality

Assessing the component quality of your solar power system is an essential first step that might impact the project's longevity and performance. High-quality components not only ensure efficient energy production but also minimize the risk of future maintenance issues and replacements. Understanding how to assess the quality of solar panels, inverters, batteries, and other system components is crucial for making informed decisions.

The process of evaluating component quality begins with researching manufacturers. Established manufacturers with a solid reputation in the industry are often synonymous with reliability and performance. Seek out businesses that have a track record of success, satisfied clients, and a commitment to solar technology advancement. Manufacturers that invest in research and development, adhere to strict quality control processes, and have their products certified by recognized industry standards are more likely to offer superior quality components.

Certifications are a crucial indicator of component quality. Solar products are tested for performance, durability, and safety in accordance with a number of international and national standards, including those established by the International Electrotechnical Commission (IEC) and the Underwriters Laboratories (UL). Components that meet or exceed these standards have undergone rigorous testing to ensure they perform under the conditions they will face once installed. Therefore, always check for certifications when evaluating the quality of solar equipment.

Another aspect of quality assessment involves examining the warranties offered by the manufacturer. Warranties not only provide peace of mind but also reflect the manufacturer's confidence in their product's durability. Components with more extended warranty periods are generally of higher quality, as the manufacturer is willing to stand behind their product for an ample time. Pay attention to what the warranty covers, including performance guarantees, which ensure that the solar panels, for instance, will not fall below a certain level of efficiency over time.

Technical specifications and performance metrics are also critical for assessing quality. For solar panels, key metrics include efficiency, temperature coefficient, and degradation rate. High-efficiency panels produce more electricity from the same amount of sunlight, while a low-temperature coefficient means the panel performs better in hot conditions.

A low degradation rate indicates the panel will retain its efficiency longer over its lifetime. Similarly, for batteries and inverters, look at capacity, efficiency, and the range of operating conditions to gauge quality.

Engaging with the solar community through forums, Social media, and online reviews can offer perceptions of the real-world performance and reliability of solar components. User experiences can highlight potential issues or advantages not apparent in product specifications or company marketing.

Lastly, consulting with solar energy professionals can offer valuable advice on component quality. Professionals who design and install solar power systems have firsthand experience with various brands and products and can recommend components based on performance, durability, and value for money.

Tips for Budget Friendly Purchases

Making budget-friendly purchases for your solar power system involves a strategic approach to ensure you're getting the best value for your investment without compromising on quality. The goal is to maximize the efficiency and longevity of your system while keeping costs manageable. Here are some tips to guide you in making cost-effective decisions for your solar project.

Researching and comparing prices is a fundamental step in the process. The solar market is competitive, and prices can vary significantly between suppliers and manufacturers. Take the time to obtain quotes from multiple sources for the components you need. This not only helps you get a sense of the price range but also positions you to negotiate better deals.

Consider the timing of your purchases. The solar industry, like many others, experiences fluctuations in pricing due to demand, new product releases, and changes in government incentives. If possible, plan your purchases to coincide with periods when prices are historically lower or when suppliers may be offering discounts to clear out inventory for new models.

Look for package deals or bundles. Suppliers often offer discounts on packages that include multiple components of a solar power system. Buying these bundles instead of each item separately can save you a lot of money. Additionally, buying components that are designed to work together can improve the overall efficiency and reliability of your system.

Don't overlook the importance of warranties and after-sales support. While it may be tempting to choose the least expensive option, the quality of customer service and the robustness of warranties can save you money in the long run. Components with more extended warranties and reputable manufacturers who stand behind their products can reduce maintenance costs and minimize downtime.

Exploring the second-hand market can also offer opportunities for budget-friendly purchases. Some companies sell refurbished or slightly used solar components at a fraction of the cost of new ones. When considering used equipment, it's essential to thoroughly assess the condition and ensure it comes with a reliable warranty.

Taking advantage of government incentives and rebates can dramatically reduce the initial cost of your solar power system. Many regions offer financial incentives for installing renewable energy systems, including tax credits, rebates, and grants. These incentives can offset a significant portion of your upfront costs, making higher-quality components more affordable.

Finally, investing in the right components from the start can lead to long-term savings. Opting for slightly more expensive but more efficient and durable components can result in lower energy bills and fewer replacements over the life of your solar power system. This approach emphasizes the total cost of ownership rather than just the initial purchase price.

By adopting a thoughtful approach to purchasing solar power system components, you can find budget-friendly options that don't sacrifice quality or performance. Research, timing, and a focus on long-term value are crucial to making cost-effective decisions that will ensure the success and sustainability of your solar energy project.

DIY vs. Professional Installation Components

Embarking on a solar power project involves critical decisions, one of which is whether to take the do-it-yourself (DIY) route or opt for professional installation. This choice affects not only the overall cost of the project but also its efficiency, reliability, and the time it takes to get your system up and running. Understanding the nuances between DIY and professional installation components can guide you in making a decision that aligns with your skills, budget, and energy goals.

DIY solar installations have gained popularity among homeowners who are handy and eager to take on new challenges. One of the primary attractions of going DIY is the potential for cost savings. By purchasing components individually and handling the installation By doing it yourself, you may save the labor expenses related to hiring a professional.

Furthermore, the DIY approach offers a deep sense of personal satisfaction and empowerment from building your own energy solution. However, it's essential to recognize that DIY projects require a significant time investment and a solid understanding of electrical systems. The responsibility for selecting compatible components, ensuring the system meets local regulations, and securing necessary permits falls entirely on you. Moreover, while the internet offers a wealth of resources, navigating technical specifications and making informed purchases can be daunting without professional guidance.

On the other hand, opting for professional installation simplifies the process considerably. Professional installers can provide a turnkey solution, from conducting site assessments and designing an optimized system to procuring high-quality components and handling the installation.

These services come with the added benefits of warranties on both the components and the workmanship. Professional installers bring expertise and experience, ensuring that your system is installed efficiently, safely, and in compliance with all local codes and regulations. While the upfront cost may be higher compared to the DIY approach, the value of having a system that is expertly installed and backed by guarantees cannot be understated. Additionally, professional installations often result in a more aesthetically pleasing setup and can maximize the system's performance and longevity, potentially offering more significant savings in the long run.

The decision between DIY and professional installation ultimately hinges on personal preferences, competencies, and financial considerations. If you have a strong background in electrical work and enjoy hands-on projects, a DIY solar installation can be a rewarding and cost-effective venture. However, if you prioritize convenience, peace of mind, and the assurance that comes with professional expertise, investing in a professionally installed system may be the better choice. Regardless of the path you choose, thoroughly researching components, understanding your energy needs, and considering the long-term implications of your decision are crucial steps in achieving a solar power system that meets your expectations and supports your journey toward sustainable energy independence.

CHAPTER 6

STEP 3 INSTALLATION PROCESS

The installation process for a solar power system transforms individual components into a unified setup that captures and converts sunlight into usable electricity. It begins with securing solar panels to the roof or another suitable location, ensuring they're optimally positioned for sunlight exposure. The inverter, which converts DC electricity from the panels into AC power, is installed next, followed by any battery storage systems for systems designed to store energy. All components are then connected through electrical wiring, adhering to safety standards and local codes. The final steps include testing and inspection to ensure the system is safely installed and fully operational. This phase is crucial and often requires technical expertise to navigate the complexities of electrical work and system integration successfully.

Safety and Preparation

Ensuring safety and thorough preparation are fundamental when embarking on the installation process of a solar power system. This phase demands attention to detail, adherence to safety protocols, and a comprehensive understanding of the current endeavor to avoid errors and guarantee the longevity and efficiency of the system. Before any physical work begins, it's essential to conduct a detailed review of the installation site, identifying any potential hazards such as unstable roofing, electrical risks, or areas prone to excessive wind or weather damage. Equipping oneself with the proper safety gear, including gloves, eye protection, and harnesses for roof work, is non-negotiable to minimize personal injury risk.

Moreover, preparation extends beyond personal safety measures. It involves a thorough review of the solar components' manuals and the local electrical codes to ensure compliance with safety standards and regulations. Gathering all necessary tools and materials before starting the installation helps streamline the process, reducing the need to pause work to source missing items. Additionally, confirming that all solar components are compatible and undamaged before installation can prevent system failures and accidents caused by faulty equipment.

In essence, prioritizing safety and preparation sets a strong base required to build a solar power system successfully. It not only safeguards those involved in the installation process but also contributes to the system's overall success, ensuring it operates safely and efficiently for years to come.

Mounting Solar Panels Techniques and Tips

For your solar power system to operate at its best and last a long time, solar panel mounting is essential. This task involves several steps, from choosing the right location to ensuring the panels are securely attached, all while maintaining safety as a top priority.

Choosing the appropriate location is the initial step. Solar panels typically perform best on south-facing roofs in the northern hemisphere, as this orientation maximizes sunlight exposure. Assessing the roof's angle, condition, and structural integrity is essential to ensure it can support the panels and allow them to operate efficiently. For less-than-ideal roof conditions, consider alternatives like ground mounts or adjustable tilt frames that can offer practical solutions.

Securely attaching the mounting hardware to the roof is a critical next step. This involves identifying the rafters or trusses to anchor the mounts firmly rather than just connecting them to the roofing material. Using a stud finder to locate rafters and creating pilot holes for the mounts can help prevent wood splitting. Incorporating flashing and sealants around the mounts will also safeguard against water leakage, preserving the roof's integrity.

When attaching the panels to the mounting hardware, it's essential to use the correct clamps or bolts and ensure there is a small gap between the roof and the panels to facilitate air circulation. This gap helps cool the panels, enhancing their efficiency. Proper spacing between the panels is also necessary to avoid shading and allow for maintenance access.

Safety considerations are paramount throughout the mounting process. Working at heights requires the use of safety gear such as harnesses, gloves, and non-slip shoes. Additionally, it's advisable to work in teams, especially when lifting and positioning the panels, to ensure they are securely fastened and to prevent accidents.

By following these techniques and tips, you can mount your solar panels safely and effectively, ensuring they deliver maximum performance over their lifespan. Whether you're tackling the project yourself or overseeing a professional installation, attention to detail and a commitment to safety will contribute to the success of your solar power system.

Electrical Wiring Basics

Electrical wiring is a fundamental component of installing a solar power system, as it connects solar panels, inverters, batteries, and other elements to create a cohesive unit that efficiently converts and distributes electricity. Understanding the basics of electrical wiring is crucial for ensuring safety, compliance with local codes, and the optimal performance of your solar system.

The electrical wiring process begins with mapping out the circuit that will connect the solar panels to the inverter and, if applicable, to the battery storage and the home's electrical panel. This circuit is crucial for the efficient transfer of electricity throughout the system. The type of wiring used must be suitable for outdoor use and rated for the voltage and current it will carry, typically involving UV-resistant and weatherproof cables to withstand environmental conditions.

Proper grounding is essential for safety and system protection. Grounding involves connecting the electrical system to the earth using grounding rods or plates, providing a path for electrical surges or faults to dissipate safely into the ground, thus protecting the system components and preventing electrical shocks.

The size of the wiring is another critical consideration. Wire gauge, or thickness, affects the amount of electricity that can safely pass through the wire without overheating. Choosing the correct wire gauge depends on the system's current and the distance the electricity must travel; longer distances and higher currents require thicker wires to minimize energy loss and heat buildup.

Color coding of wires is a standard practice to identify their purpose and ensure correct connections. Typically, black or red-wires are used for positive connections, white or grey for neutral, and green or bare copper for grounding. Adhering to these conventions can prevent wiring errors and facilitate troubleshooting and maintenance.

Connectors play a vital role in maintaining the integrity of electrical connections. They must be compatible with the type of wiring and designed for outdoor use to prevent moisture ingress and corrosion. Secure, watertight connections are essential for long-term reliability and safety.

Finally, installing a solar power system involves compliance with local electrical codes and standards, which may require specific types of wiring, protective devices such as circuit breakers and fuses, and inspection by a certified electrician. These regulations are designed to ensure the safety and efficiency of electrical installations.

Understanding the basics of electrical wiring in the context of solar power systems is crucial for anyone involved in the installation process. Whether you are a DIY enthusiast or working with professional installers, knowledge of wiring principles, safety practices, and local codes is essential for creating a safe, efficient, and compliant solar energy system.

Installing the Charge Controller and Inverter

The installation of the charge controller and inverter is a pivotal step in setting up a solar power system, as these components play critical roles in managing energy flow and converting solar power into usable electricity. In order to avoid overcharging and increase battery life, the charge controller controls the batteries' voltage and current when they enter the solar panels. The inverter, on the other hand, transforms the direct current (DC) from the solar panels or batteries into alternating current (AC), the form of electricity compatible with household appliances and the grid.

When installing the charge controller, it's essential to position it between the solar panels and the batteries to monitor and control the power flow to the batteries. The location should be relatively close to the batteries to minimize voltage drop but in a well-ventilated area to dissipate heat generated during operation. The controller must be securely mounted on a non-conductive surface to avoid electrical shorts and allow for easy access for monitoring and maintenance.

The wiring connections from the solar panels to the charge controller and from the controller to the batteries must adhere to the manufacturer's specifications regarding wire gauge and type. Proper fusing is also critical to protect against overcurrent situations. It's vital to follow the recommended sequence of connections, usually connecting the batteries to the charge controller first, to prevent the controller from operating without a load, which can cause damage.

Installing the inverter involves choosing a location that is close to the battery bank to reduce energy loss in the wiring but also in a place that is dry, cool, and well-ventilated to prevent overheating. The inverter should be mounted securely to a wall or other stable surface, ensuring that it is level and that there's ample space around it for airflow and easy access.

The electrical connections for the inverter include connecting it to the battery bank with the appropriate gauge wiring and installing a disconnect switch between them for safety. The AC output side of the inverter will connect to your home's electrical panel or to specific appliances, depending on your system configuration. Grounding the inverter is also essential for safety, ensuring that any fault currents are appropriately directed to the earth to prevent electric shock.

Finally, both the charge controller and the inverter should be configured according to your system's specifications and the manufacturer's guidelines. This might involve setting charge parameters on the controller to match the battery type and capacity and configuring the inverter for the correct voltage and frequency of your electrical system.

Proper installation of the charge controller and inverter is crucial for the safety, efficiency, and longevity of your solar power system. While many aspects of their installation can be handled by individuals with a good understanding of electrical systems, consulting with or hiring a professional can ensure that these components are installed correctly and in compliance with local electrical codes.

Setting Up the Battery Bank

Setting up the battery bank in a solar power system is a crucial step that enables energy storage for later use, ensuring a continuous power supply during the night or on cloudy days. This process involves selecting the correct type of batteries, configuring them correctly, and installing them safely.

The first consideration in setting up a battery bank is choosing the appropriate battery type for your needs. Lead-acid batteries, including flooded lead-acid and sealed lead-acid (AGM and gel), have been traditionally used due to their cost-effectiveness and availability. However, lithium-ion batteries are increasingly popular due to their higher efficiency, longer lifespan, and lower maintenance requirements, albeit at a higher initial cost. The choice between these types depends on your budget, energy needs, and maintenance capacity.

Once the battery type is selected, determining the size of the battery bank is essential. This involves calculating your energy consumption needs and considering the autonomy of your system — how long you need the system to supply power without sunlight. The capacity of the battery bank should be large enough to store sufficient energy for your needs while avoiding deep discharges, which can shorten battery life.

The configuration of the battery bank is another critical aspect. Series battery connections can raise the voltage to meet system needs or in parallel to increase capacity (ampere-hours, Ah) for a more extended energy supply. Higher voltage and capacity can also be attained by combining parallel and series connections. It's crucial to use batteries of the same type, age, and capacity to ensure balanced charging and discharging.

Installing the battery bank involves placing the batteries in a well-ventilated area to prevent the accumulation of gases and heat, which can be hazardous. For lead-acid batteries, especially flooded types, this is particularly important due to the off-gassing that occurs during charging. The area should be clean, dry, and protected from extreme temperatures to maintain battery efficiency and lifespan. Additionally, securing the batteries in place will prevent movement that could lead to loose connections or damage.

Safety is paramount when setting up a battery bank. This includes installing a battery management system (BMS) for lithium-ion batteries, which monitors and manages the battery's operation to prevent overcharging, deep discharging, and overheating. Protective measures such as fuses or circuit breakers should be incorporated to protect against overcurrent situations. Proper grounding of the battery bank is also essential to safeguard against electrical shocks.

Finally, the battery bank should be connected to the charge controller and inverter with the correct gauge and type of wiring to handle the expected current flow. Ensuring tight and secure connections can prevent energy loss and potential fire hazards.

Setting up the battery bank is a meticulous process that lays the foundation for a reliable and efficient solar power system. While individuals with a solid understanding of electrical systems can undertake this task, consulting with or hiring a professional can ensure that the battery bank is configured and installed correctly, adhering to safety standards and local regulations.

System Grounding and Protection

System grounding and protection are essential components of a safe and effective solar power system. Grounding serves to protect people from electric shock, safeguard equipment from lightning strikes and electrical surges, and ensure the overall electrical safety of the installation. Protection mechanisms, such as fuses and circuit breakers, further secure the system against over currents and short circuits, preventing damage to components and reducing fire risks.

Grounding involves creating a direct physical connection between the electrical system and the earth. This connection provides a path for fault currents to flow safely into the ground, significantly reducing the risk of electric shock to individuals and preventing potential damage to the solar power system during electrical faults or lightning events. The grounding process typically includes grounding the metal frames of solar panels, the inverter, and other metallic components, ensuring that any stray currents are immediately directed away from the system and its users.

To establish a robust grounding system, conductive materials such as copper rods or plates are driven into the ground, and electrical cables connect these grounding points to the solar power system's components. It's crucial that these connections are secure and made using appropriate conductive materials to ensure practical grounding over the life of the system.

Protection against over currents and short circuits is achieved through the strategic placement of fuses and circuit breakers within the solar power system. Fuses are designed to break the circuit by melting when the current flow exceeds a certain threshold, preventing excessive current from damaging the system. Circuit breakers, on the other hand, automatically interrupt the flow of electricity when they detect an overload or short circuit, offering a reusable solution to manage system safety.

Both fuses and circuit breakers should be selected based on the maximum expected current in the circuit they protect and should be installed at critical points in the system, including the connections between solar panels and the charge controller, the charge controller and the battery bank, and the battery bank and the inverter. This ensures that every component of the solar power system is safeguarded against electrical mishaps.

Additionally, surge protection devices (SPDs) are used to protect the system from voltage spikes often caused by lightning or grid fluctuations. SPDs absorb or redirect excess voltage, preventing it from traveling through the system and causing damage.

Implementing effective system grounding and protection is a fundamental aspect of solar power system design, ensuring safety and reliability. Adhering to local electrical codes and standards and possibly consulting with a professional can guide the proper selection and installation of grounding and protection components. This meticulous approach not only safeguards the investment in solar power but also ensures the safety of individuals and property.

CHAPTER 7

STEP 4 TESTING AND MAINTENANCE

Initial System Testing and Troubleshooting

After installing a solar power system, conducting initial system testing and troubleshooting is essential to confirm that every part is operating appropriately and effectively. This phase helps identify any installation errors or component malfunctions early on, allowing for adjustments or repairs before the system is fully operational.

The process begins with a visual inspection of the entire solar setup to confirm that all components are securely mounted, correctly oriented, and properly connected. This includes checking the solar panels, inverter, charge controller, batteries, and all wiring and connections for any visible signs of damage or incorrect installation.

Following the visual inspection, electrical testing is conducted to verify the system's electrical integrity and performance. This involves using a multimeter or a solar-specific meter to measure voltage and current output from the solar panels to ensure they match the expected values. Checking the voltage at various points, such as at the panel outputs, the input and output of the charge controller, and the inverter, can reveal any discrepancies that might indicate wiring issues or component failures.

Testing the charge controller involves verifying that it correctly regulates the voltage and current flowing to and from the battery bank, protecting the batteries from overcharging or excessive discharge. Similarly, the inverter's output is tested to ensure it is correctly converting DC electricity from the panels or batteries into AC electricity at the proper voltage and frequency for household use or grid connection.

Battery testing is also crucial, particularly for systems that include energy storage. This can involve checking each battery's voltage and state of charge to ensure they are within normal ranges. Load testing may also be performed to assess the battery bank's ability to hold and deliver power under typical usage conditions.

If any issues are identified during the initial testing phase, troubleshooting begins to isolate and resolve the problems. This might involve rechecking all connections, replacing faulty wiring, adjusting the configuration of solar panels, recalibrating the charge controller, or resetting the inverter. In cases where hardware defects are suspected, component replacement may be necessary.

For complex issues, consulting the manufacturer's documentation or technical support can provide guidance and troubleshooting steps specific to the components in question. Additionally, leveraging online forums or contacting a professional solar technician can offer insights and solutions based on similar experiences.

Initial system testing and troubleshooting are critical steps to ensure the solar power system is safe, compliant, and ready for operation. This proactive approach not only safeguards the investment in solar energy but also optimizes the system's performance and efficiency, paving the way for reliable renewable energy production.

Routine System Maintenance

Routine system maintenance is key to ensuring the long-term efficiency and reliability of a solar power system. Regular maintenance activities help identify and address potential issues before they escalate into costly repairs, ensuring the system operates at optimal performance throughout its lifespan.

The cornerstone of routine maintenance is the visual inspection of all solar panels and system components. This involves checking for any physical damage, such as cracks in the panels, corrosion on the frames or wiring, and loose or disconnected cables. Special attention should be paid to the mounting hardware to ensure that the panels remain securely fastened, especially after severe weather events.

Cleaning the solar panels is another critical maintenance task. Dust, dirt, bird droppings, and fallen leaves can accumulate on the panel surfaces, blocking sunlight and reducing the system's energy output. Regular cleaning with water and a soft brush or cloth can remove these obstructions, restoring the panels' efficiency. The frequency of cleaning will depend on the location and environment, with some sites requiring more frequent attention than others.

Checking the system's electrical components is also essential. This includes verifying that all electrical connections are tight and free from corrosion, which can cause resistance and reduce the efficiency of energy transfer. The charge controller and inverter should be inspected for any error indicators or malfunction signs, and their settings should be reviewed to ensure they are still optimal for the system's current operating conditions.

For systems with battery storage, battery maintenance is crucial. For lead-acid batteries, this may involve checking the electrolyte levels and topping up with distilled water as needed, as well as ensuring the terminals are clean and corrosion-free. For all battery types, monitoring the state of charge and overall health of the batteries can help detect any issues early, extending the battery life and preventing system downtime.

Finally, maintaining accurate records of all maintenance activities, performance data, and any issues encountered can provide valuable insights into the system's operation over time. This historical data can help identify trends, such as decreasing efficiency, that may indicate the need for component replacement or system upgrades.

Routine system maintenance, while straightforward, is essential to the efficient running of a solar power system. By dedicating time and attention to regular inspections, cleaning, and component checks, Owners of systems may increase the amount of energy they produce, prolong the life of their system, and guarantee a consistent flow of clean, renewable energy.

Monitoring and Optimizing Performance

Monitoring and optimizing the performance of a solar power system are critical steps to ensure it operates efficiently and effectively over its lifespan. This process involves tracking the system's output, identifying areas for improvement, and making adjustments to enhance energy production and overall system health.

Utilizing system monitoring tools and software, which may offer real-time data on energy generation, consumption, and system performance parameters, is the cornerstone of performance monitoring. Many modern inverters and charge controllers come with built-in monitoring capabilities, accessible via smartphone apps or web portals. These tools allow system owners to track daily, monthly, and annual energy production, compare it against expected performance and identify any discrepancies that may indicate issues.

Monitoring also extends to environmental factors that can impact system performance, such as shading from trees or buildings that may have changed over time, accumulation of dirt or debris on the panels, and weather patterns. Regularly comparing the system's output against similar systems in the same geographic area can help identify potential underperformance issues.

Optimizing system performance often starts with addressing any identified issues, such as cleaning solar panels to remove obstructions or adjusting the angle or orientation of panels to better capture sunlight. For systems experiencing significant shading, trimming or removing obstructions can significantly improve performance.

For systems with battery storage, optimization includes ensuring that the batteries are being charged and discharged within their recommended parameters to maximize lifespan and efficiency. This may involve adjusting charge controller settings based on seasonal changes in sunlight availability or usage patterns.

Software updates are another aspect of optimization. Manufacturers regularly release firmware updates for inverters, charge controllers, and monitoring systems that can improve functionality, introduce new features, or enhance system efficiency. Keeping these components updated can ensure your system benefits from the latest technological advancements.

In addition to hardware adjustments, optimizing performance can also involve reevaluating energy consumption patterns within the home or business. Putting energy-saving measures into practice, including updating to LED lighting, using energy-efficient appliances, and improving insulation, can reduce overall energy demand and increase the net benefit of the solar power system.

Ultimately, monitoring and optimizing the performance of a solar power system is an ongoing process that requires attention and action from the system owner. By staying engaged with the system's operation, addressing issues promptly, and making adjustments as necessary, system owners can ensure their solar power system continues to provide reliable, efficient energy production throughout its operational life.

Upgrades and System Expansion

As the needs of a household or business evolve, or as advancements in solar technology continue to emerge, upgrading and expanding a solar power system becomes a consideration for many system owners. This process not only enhances the system's capacity and efficiency but also adapts to changing energy demands, ensuring the solar installation continues to meet the owner's needs effectively.

Upgrading a solar power system can involve several approaches, including adding more solar panels to increase the system's energy production capacity. This is particularly relevant for those who have experienced an increase in energy consumption or who wish to capitalize on unused roof space. When adding panels, it's essential to ensure they are compatible with the existing system components, such as the inverter and charge controller, or consider upgrading these components as well to handle the increased output.

Another upgrade path is replacing older solar panels or components with newer, more efficient models. Solar technology has seen significant advancements in efficiency and reliability, and upgrading can result in better performance and energy production from the same footprint. This might include investing in panels with higher efficiency ratings, more robust inverters with enhanced features, or battery storage systems with greater capacity and longevity.

System expansion often goes hand-in-hand with upgrades. Expanding a solar power system might not only involve adding more panels but also integrating energy storage if it wasn't part of the original setup. Battery storage can significantly enhance a system's utility by reserving extra energy for use during periods of high demand. Times or when the sun isn't shining, decreasing dependency on the grid and boosting energy independence.

Before proceeding with upgrades or expansion, a thorough evaluation of the existing system and energy needs is crucial. This evaluation should consider the physical space available for additional panels, the electrical capacity of the current system, and the potential need for additional or upgraded components to support the expanded system. It's also important to review local regulations and utility policies, as they may have changed since the original installation and could impact your expansion plans.

Financial incentives and rebates should also be revisited, as new programs may be available to help offset the cost of upgrading and expanding your solar power system. Furthermore, speaking with an expert in solar energy may offer insightful advice on the best and most affordable upgrade options, guaranteeing that the enlarged system maximizes return on investment.

Upgrades and system expansion offer a pathway to not only meet increased energy demands but also leverage improvements in solar technology for greater efficiency and performance. By carefully planning these enhancements, system owners can ensure their solar power systems continue to provide sustainable, clean energy tailored to their evolving needs.

CHAPTER 8

LIVING OFF THE GRID

Living off the grid with a solar power system represents a profound shift towards energy independence and efficiency as well as a more powerful connection with the environment. This lifestyle choice involves relying on your solar installation, possibly in combination with other renewable energy sources, to meet all your electricity needs without the support of the public utility grid. It's a journey that requires careful planning, a commitment to energy efficiency, and an embrace of innovative solutions to energy challenges.

The foundation of off-grid living is a well-designed solar power system tailored to your specific energy needs. This involves accurately calculating your daily and seasonal energy consumption and designing a system with enough capacity to meet these demands. Because off-grid systems must operate independently, they often include a significant battery storage component to ensure electricity is available at night and during periods of low sunlight. Additionally, incorporating redundancy, such as backup generators or wind turbines, can enhance reliability.

Energy efficiency becomes paramount when living off the grid

By reducing overall energy consumption, you can minimize the size and cost of the solar power system needed. This often involves using energy-efficient appliances, LED lighting, and adopting practices that reduce electricity use. Designing your home for passive solar gain, excellent insulation, and natural cooling can also significantly reduce heating and cooling demands, further decreasing the need for generated electricity.

Adapting to the rhythms of nature is a key aspect of off-grid living. This might mean adjusting energy-intensive activities to times of peak solar production or being more mindful of energy use during periods of limited sunlight. It's a lifestyle that fosters a deep awareness of your energy consumption and its direct link to the environment.

Managing water supply and waste disposal are also critical components of living off the grid, often involving rainwater collection systems and septic tanks or composting toilets. These systems complement your solar installation, extending the principles of sustainability and self-sufficiency to other aspects of daily living.

Living off the grid with a solar power system is not just about the technical setup but also about embracing a philosophy of life that values self-reliance, environmental stewardship, and a simpler, more intentional way of living. It's a rewarding journey that connects you more closely with the natural world, providing a profound sense of achievement and peace from knowing you are living in harmony with the environment. While the transition requires significant preparation and adaptation, the benefits of off-grid living, from the independence it grants to the positive impact on the planet, make it a deeply fulfilling choice for many.

Daily Life with Solar Power

Embracing solar power transforms daily life in ways that resonate with sustainability, efficiency, and a deeper connection to the natural environment. This shift to renewable energy affects various aspects of household management, energy consumption patterns, and even the financial dynamics of utility use. Living with solar power means adapting to the rhythms of the sun, becoming more energy-conscious, and enjoying the benefits of clean, renewable energy.

One of the most immediate impacts of solar power on daily life is the reduction in electricity bills. Solar panels generate electricity during daylight hours, reducing or even eliminating the need for electricity from the grid and thereby significantly lowering monthly energy costs. For homes with net metering, excess electricity produced can be sent back to the grid, potentially earning credits and further offsetting costs during less sunny periods or overnight.

The daily management of a solar-powered home involves monitoring energy production and consumption. Many modern solar systems come equipped with monitoring software that provides real-time data on energy generation and usage. This information allows residents to make informed decisions about when to run energy-intensive appliances, such as dishwashers, washing machines, and electric vehicle chargers, to coincide with peak production times, maximizing the use of solar energy.

Energy efficiency becomes more than a cost-saving strategy; it's a lifestyle choice. Households with solar power often invest in energy-efficient appliances, LED lighting, and smart home technologies that automate energy savings. These measures not only reduce overall consumption but also enhance the effectiveness of the solar power system, ensuring that energy needs are comfortably met within the system's capacity.

Living with solar power also fosters a heightened awareness of the environment and the importance of sustainable living. It encourages a shift in behavior towards conserving energy, reducing waste, and minimizing one's carbon footprint. This can extend beyond electricity use to include water conservation, recycling, and choosing sustainable products and services.

For those with battery storage systems, daily life includes the added dimension of energy management and storage. Batteries store solar energy collected during the day for use at night or during cloudy days, providing a continuous power supply. Managing the battery system to ensure it meets energy needs while maintaining the health and longevity of the batteries becomes part of the routine.

Moreover, daily life with solar power brings a sense of empowerment and independence. Generating your own electricity from the sun reduces reliance on utility companies and the grid, providing a buffer against rising energy prices and supply disruptions. It also contributes to a cleaner, more sustainable world by reducing dependence on fossil fuels and lowering greenhouse gas emissions.

Energy Conservation Tips

Adopting energy conservation practices is essential for enhancing the efficiency of your home and making the most of your solar power system. By reducing unnecessary energy consumption, you not only lower your utility bills but also contribute to a more sustainable environment. Here are actionable energy conservation tips that can lead to significant savings and a reduced carbon footprint.

Leverage Natural Light: Make the most of natural daylight to illuminate your home, reducing the need for artificial lighting. Consider rearranging your workspace to be near windows and installing skylights or light tubes in darker areas.

Invest in Energy-Efficient Appliances: When it's time for new appliances, choose those with the best energy efficiency ratings. Energy Star-rated appliances, for instance, consume less electricity and perform as effectively as their less efficient counterparts.

Adopt Smart Thermostat Technology: Smart thermostats automatically adjust your home's temperature based on your habits and preferences, ensuring energy isn't wasted on heating or cooling when not needed.

Switch to LED Lighting: LEDs are far more energy-efficient than traditional incandescent bulbs, using up to 90% less power for the same amount of light output. They also last longer, reducing the need for frequent replacements.

Insulate Your Home: Proper insulation helps maintain your home's temperature, reducing the energy required for heating and cooling. Check and upgrade insulation in your walls, roof, and floors as needed, and seal any drafts around doors and windows.

Use Smart Power Solutions: Phantom power drain from electronics that remain plugged in can add up. Utilize smart power strips that cut off power when devices are not in use, eliminating this hidden energy consumption.

Optimize Water Heating: Water heating is a significant energy user. Lower your water heater's temperature setting to around 120 degrees Fahrenheit and insulate the tank to keep water hot without excessive energy use.

Cook Efficiently: Small changes in the kitchen, like using lids on pots to speed up cooking and opting for the microwave over the oven when possible, can reduce energy usage. Avoid opening the oven frequently to check on food, as this lowers the temperature and requires more energy to heat back up.

Embrace Passive Cooling and Heating: Use thermal drapes or blinds to control the amount of sunlight entering your home, cooling it naturally in the summer and capturing warmth in the winter. Strategic landscaping can also provide shade and act as a windbreak.

Regularly Monitor Energy Use: Keeping an eye on your energy consumption can help identify areas for improvement. Many utilities offer energy monitors, and solar systems often include monitoring apps that detail your production and usage.

Implementing these tips into your daily routine can significantly impact your energy consumption and solar power system's effectiveness. Energy conservation not only supports environmental sustainability but also enhances the autonomy and financial benefits of living with solar power.

Dealing with Power Outages

Incorporating strategies to deal with power outages is an essential aspect of planning and living with a solar power system, especially for those relying heavily on solar energy for their daily needs. While solar power systems are reliable, external factors like severe weather can lead to disruptions in energy supply. Here's how you can prepare for and deal with power outages effectively.

Firstly, consider integrating a battery storage system into your solar setup if you haven't already. Batteries can store excess energy generated during the day for use at night or during an outage, providing a continuous power supply when the grid is down. The capacity of your battery storage should be tailored to meet your critical energy needs for the duration of typical outages in your area.

Another effective strategy is to install a backup generator. While solar power systems with battery storage can cover basic needs, a generator can ensure that your home remains fully functional during prolonged outages. Generators can run on various fuels, including gasoline, diesel, or propane. Choosing a generator that integrates seamlessly with your solar system requires careful planning and, often, professional advice.

Investing in a hybrid inverter can also enhance your resilience to power outages. Hybrid inverters allow solar systems to operate independently of the grid and can manage the flow of electricity from solar panels, the battery, and a generator, ensuring that power is available when needed. Some models can automatically switch to off-grid mode during an outage, providing an uninterrupted power supply.

For systems already equipped with battery storage, regularly testing and maintaining your batteries ensures they are ready to perform when needed. This includes monitoring charge levels, ensuring proper operation of the battery management system, and keeping connections clean and tight.

Understanding your energy priorities is crucial in managing consumption during an outage. Identifying essential loads, such as refrigeration, lighting, and medical devices, allows you to allocate power more effectively, ensuring that critical needs are met. Some systems allow for load prioritization, directing power to essential services automatically.

Preventive measures, such as ensuring your solar panels and components are well-maintained, can also reduce the risk of system failures that might coincide with grid outages. Regular inspections and cleaning of panels, along with checks for wear and tear on wiring and connections, contribute to the overall reliability of your solar power system.

Finally, staying informed about potential power outages through local news or weather apps can help you prepare in advance. Charging battery storage to full capacity and reducing energy consumption as a storm -

approaches can extend the duration your system can provide power during an outage.

Dealing with power outages effectively requires a combination of system preparedness, strategic planning, and an understanding of your energy needs. By adopting these strategies, you can ensure that your reliance on solar power remains a reliable and sustainable choice, even in the face of unexpected disruptions.

Long-Term Sustainability

Achieving long-term sustainability with a solar power system involves more than just the initial switch to renewable energy. It encompasses a commitment to ongoing maintenance, environmental stewardship, and adapting to evolving energy needs to ensure that the system continues to operate efficiently and effectively over its lifespan. This approach not only maximizes the environmental benefits of solar power but also ensures the system remains a viable and cost-effective energy solution for years to come.

One of the key aspects of long-term sustainability is regular maintenance and inspection of the solar power system. Ensuring that solar panels are clean and free from obstructions, checking for and repairing any damage, and replacing worn-out components can prevent performance degradation and extend the system's useful life. Additionally, keeping inverters, batteries, and other system components in good working order through regular checks and updates can optimize efficiency and prevent costly repairs.

Another crucial factor is the adaptation to changing energy needs. Over time, a household or business's energy consumption can change due to factors such as growth, the adoption of new technologies, or increased efficiency. Periodically reassessing your energy needs and adjusting the solar power system accordingly—whether by expanding capacity, upgrading components, or integrating new technologies—can ensure that the system continues to meet these evolving demands.

Technological advancements play a significant role in long-term sustainability. The solar power industry is rapidly evolving, with new innovations that offer improved efficiency, reliability, and integration with other renewable energy sources. Staying informed about these developments and considering upgrades or additions to the system can leverage these advancements to enhance performance and sustainability.

Environmental stewardship is also integral to the long-term sustainability of a solar power system. This includes responsible management of system components at the end of their lifecycle, such as recycling solar panels and batteries, to minimize environmental impact. Additionally, supporting and participating in broader sustainability practices and renewable energy initiatives can amplify the positive environmental effects of individual solar power systems.

Finally, education and advocacy are powerful tools for promoting long-term sustainability. Sharing experiences, knowledge, and the benefits of solar power with the community can inspire others to consider renewable energy solutions. Advocacy for policies and programs that support renewable energy development and integration can also contribute to a more sustainable and resilient energy future.

Long-term sustainability with a solar power system is an ongoing journey that requires attention, adaptation, and a proactive approach to maintenance and environmental responsibility. By embracing these principles, solar power system owners can ensure their investment continues to provide clean, renewable energy while contributing to a healthier planet for future generations.

How You Can Share Your Review Through Amazon.com

- Visit the Amazon page where you purchased or found my book.
- Scroll down to the 'Customer Reviews' section near the bottom of the page.
- Click on 'Write a customer review' to begin sharing your valuable insights and experiences with the book.

Instant QR Code Access

Simply scan the QR code below with your smartphone to be directly taken to the Amazon review section for the book. This quick access method makes it easy for you to leave your feedback without navigating through the website.

Your review is not just feedback for us; it's a beacon for future readers navigating the world of solar energy and sustainable living. Thank you for taking the time to share your thoughts and helping us spread the message of sustainability.

CHAPTER 9

BEYOND THE BASICS

Exploring beyond the basics of solar power involves delving into advanced technologies, innovative practices, and broader applications that enhance the functionality, efficiency, and impact of solar energy systems. As the world moves towards a more sustainable and renewable energy-driven future, pushing the boundaries of solar power can lead to significant advancements in how we generate, store, and use energy.

One area of exploration is the integration of solar power with smart home technologies. Intelligent energy systems allow for more precise control and optimization of energy use, enabling homeowners to manage their solar power generation, storage, and consumption in real-time. This can include automating energy-intensive tasks to coincide with peak production hours, dynamically adjusting to energy demands, and even participating in grid services through demand response programs.

Another frontier is the development and adoption of battery storage technologies. Advances in battery storage not only improve the efficiency and lifespan of energy storage solutions but also make solar power more reliable and accessible, even in off-grid or unstable grid situations. Emerging technologies like solid-state batteries and flow batteries offer promising improvements in safety, capacity, and sustainability over traditional battery types.

The combination of solar power with other renewable energy sources, such as wind or hydroelectric power, can create more robust and resilient energy systems. Hybrid systems can ensure a continuous power supply by compensating for the variability of solar energy with other renewables that may have different production patterns. This holistic approach to renewable energy harnesses the strengths of each source, moving closer to a fully sustainable energy solution.

Community solar projects represent another avenue for expanding the reach and benefits of solar power. These projects allow individuals who may need more space or means to install their own solar panels to invest in solar energy collectively. By pooling resources, community solar projects make renewable energy more accessible and equitable, spreading its environmental and economic benefits more widely.

Advancements in solar panel technology also play a crucial role in moving beyond the basics. Research into new materials and designs, such as perovskite solar cells and bifacial solar panels, aims to increase efficiency and reduce the cost of solar panels. Such innovations could revolutionize solar power, making it more affordable and efficient and opening up new applications for solar energy.

Finally, exploring solar power beyond the basics involves advocating for and implementing policies that support the growth and integration of renewable energy. This includes incentives for renewable energy investments, regulations that support grid integration of distributed energy resources, and initiatives that encourage research and development in renewable technologies.

As we venture beyond the basics of solar power, the potential for innovation and transformation in how we generate and use energy is immense. By embracing advanced technologies, integrating with other renewable sources, and fostering community and policy support, solar power can play a pivotal role in achieving a sustainable, clean energy future.

Hybrid Systems Incorporating Wind and Hydro

Hybrid energy systems that combine solar power with wind and hydroelectric sources represent a comprehensive approach to renewable energy, offering a more reliable and consistent power supply. By harnessing the strengths of each energy source, these systems can provide electricity in a wide range of environments and conditions, mitigating the variability associated with individual renewable resources.

Integrating Solar and Wind Power: Solar and wind energy complement each other well because of their contrasting generation patterns. Solar panels produce the most energy during sunny days, while wind turbines can generate power at night and during cloudy, windy conditions. This complementary relationship means that when solar power generation decreases, wind power can often compensate, and vice versa, leading to a more stable and continuous energy supply. This synergy not only enhances reliability but also increases the overall efficiency of the energy system, making it particularly suitable for off-grid applications or regions with limited grid infrastructure.

Incorporating Hydroelectric Power: Small-scale hydroelectric systems, or micro-hydro, can be an excellent addition to solar and wind energy systems, especially in areas with suitable water resources. Unlike solar and wind, hydroelectric power can provide a constant energy output unaffected by daily or seasonal changes in weather. When integrated into a hybrid system, hydroelectric power can offer a steady base load of electricity, with solar and wind sources providing additional power as available. This setup is especially beneficial in regions with varying water flow rates throughout the year, as solar and wind can offset decreases in hydroelectric output during dryer periods.

Design and Implementation Challenges: Creating an effective hybrid system requires careful planning and design to balance the capacities of solar, wind, and hydro components according to local climate and geographical conditions. It involves sophisticated control systems capable of managing the variable inputs from each energy source and optimizing power distribution and storage. Additionally, considerations such as environmental impact assessments, regulatory approvals, and community engagement are crucial for the successful implementation of hybrid systems, especially when incorporating hydroelectric power.

Energy Storage and Management: The inclusion of energy storage solutions, such as batteries or pumped hydro storage, is critical in hybrid systems to accommodate the excess energy produced and ensure a consistent supply during periods of low generation. Advanced energy management systems can dynamically adjust the operation of each component based on real-time data and energy demand, further enhancing the system's efficiency and resilience.

Advantages of Hybrid Systems: Hybrid systems incorporating solar, wind, and hydropower offer several benefits, including reduced dependence on fossil fuels, decreased carbon footprint, and enhanced energy security. By diversifying energy sources, these systems can adapt to changing environmental conditions and energy demands, providing sustainable and reliable power to communities, industries, and remote locations.

Hybrid renewable energy systems exemplify the innovation and adaptability required to transition to a more sustainable energy future. By combining solar, wind, and hydroelectric power, these systems leverage the unique benefits of each source, offering a versatile and robust solution to meet the world's growing energy needs.

Solar-Powered Water Heating

Solar-powered water heating is an efficient and sustainable method to harness the sun's energy for domestic and commercial hot water needs. This technology captures solar radiation to heat water, significantly reducing reliance on conventional energy sources like electricity or natural gas, leading to lower energy bills and a reduced carbon footprint.

The core component of a solar water heating system is the solar collector, typically installed on the roof or in an area with significant sun exposure. There are two main types of solar collectors used in water heating systems: flat-plate collectors and evacuated tube collectors. Flat-plate collectors, consisting of a dark flat-plate absorber, a transparent cover, and a heat-insulating backing, are widely used for their durability and cost-effectiveness. Evacuated tube collectors, which consist of rows of parallel, fine glass tubes, offer higher efficiency, especially in cooler climates, by minimizing heat loss.

The heated water from the solar collectors is then stored in an insulated water tank, similar to conventional water heaters, allowing for the availability of hot water even when the sun is not shining. Some systems use a direct circulation method, where water is heated as it passes directly through the solar collectors and then stored in the tank. Others employ an indirect circulation system, which uses a heat-transfer fluid to capture heat from the solar collectors and then transfer it to the water in the storage tank via a heat exchanger. Indirect systems are particularly beneficial in areas prone to freezing temperatures, as they prevent the water in the collectors from freezing.

For optimal performance, solar water heating systems may be integrated with a backup water heating source, ensuring hot water availability during periods of low sunlight, such as on cloudy days or during nighttime. This backup system can be powered by conventional energy sources but is used minimally, as the solar system provides the majority of the heating energy required.

Implementing a solar-powered water heating system involves considerations such as the size of the system, which should be based on the water heating needs of the household or facility and the local climate, as solar insolation levels can significantly impact system performance. Additionally, the orientation and tilt of the solar collectors are crucial factors that affect the system's efficiency and should be optimized to maximize sun exposure.

Solar water heating systems not only offer environmental benefits by reducing greenhouse gas emissions but also provide financial savings over time through reduced energy costs. With advancements in technology and increasing accessibility, solar-powered water heating has become a viable and attractive option for sustainable energy management in residential, commercial, and industrial applications. By leveraging the abundant and free energy from the sun, these systems exemplify a practical approach to reducing our reliance on fossil fuels and moving towards a more sustainable energy future.

Automating Your Solar Power System

Automating your solar power system represents a significant step towards optimizing energy efficiency, enhancing convenience, and maximizing the return on your investment in solar technology. Automation involves integrating intelligent technologies and control systems into your solar setup to manage and monitor energy production, storage, and consumption seamlessly and in real-time.

The heart of solar system automation is an intelligent controller or energy management system that interfaces with the solar panels, inverter, battery storage, and, in some cases, home energy appliances. These systems use sophisticated algorithms to analyze patterns in energy generation and usage, adjusting the flow of electricity to where it's needed most and when it's needed, ensuring optimal use of the solar energy produced.

One key aspect of automating your solar power system is the dynamic management of energy storage. Intelligent controllers can decide the best times to store energy in batteries, taking into account factors like the forecasted weather, expected energy consumption, and current energy prices if connected to the grid. This ensures that excess solar power is stored efficiently and made available during peak demand times or when solar production is low, such as during the evening or on cloudy days.

Another benefit of automation is the ability to integrate with smart home systems and devices. This includes smart thermostats, lighting, and appliances that can be programmed or controlled remotely to use energy more efficiently. For example, non-essential devices can be automatically turned off during periods of low solar production, while essential functions like heating and cooling can be optimized to match solar energy availability, enhancing overall system efficiency.

Automating your solar power system also provides enhanced monitoring capabilities. Through user-friendly apps and web interfaces, homeowners can receive detailed insights into their system's performance, including real-time data on energy production, battery levels, and consumption patterns. This level of visibility allows for immediate adjustments, proactive maintenance, and informed decision-making to improve system performance and longevity.

Additionally, automation can extend to predictive maintenance and troubleshooting, where the system can alert homeowners to potential issues before they become significant problems. This might include notifications about underperforming solar panels, battery health, or system inefficiencies, enabling timely interventions that maintain the system's optimal operation.

Investing in automation technologies not only simplifies the management of your solar power system but also significantly increases its efficiency and effectiveness. By leveraging the power of intelligent technologies, homeowners can enjoy a more resilient, responsive, and user-friendly solar energy experience, ensuring that their renewable energy system continues to meet their needs and exceed their expectations well into the future.

Community Solar Projects and Sharing

Community solar projects represent an innovative approach to renewable energy, allowing individuals who may not have the means or the space to install their own solar panels to still benefit from solar power. These projects democratize access to solar energy, fostering broader community participation in the transition to renewable sources. By pooling resources and sharing the benefits of a single, large-scale solar installation, community members can reduce their energy bills, support local job creation, and contribute to the reduction of greenhouse gas emissions.

Community solar projects typically involve a solar array installed on a shared community space, such as a public building's roof, a brownfield site, or unused land. Participants can subscribe to the solar project, purchasing a share of the power produced, which then offsets their own electricity consumption and costs. This model is particularly appealing for renters, those with shaded or unsuitable roofs, and individuals in multi-tenant buildings, offering them a stake in renewable energy generation without the need for on-site installation.

The benefits of community solar are multifaceted. Financially, participants can enjoy reduced electricity costs over time, as the price of solar energy from the project is often lower than the retail rate from the grid. Environmentally, community solar contributes to the reduction of carbon footprints, harnessing clean, renewable energy and decreasing reliance on fossil fuels. Socially, these projects foster a sense of community ownership and engagement in renewable energy initiatives, raising awareness about the importance of sustainability and environmental stewardship.

Implementing a community solar project involves collaboration between community members, local governments, utility companies, and solar developers. Key considerations include securing a suitable location for the solar array, determining the project's size based on community demand, and navigating regulatory and financial frameworks to ensure the project's viability and accessibility. Policies such as virtual net metering, which allows for the credit from solar production to be shared among participants, are crucial for the equitable distribution of benefits.

For those interested in participating in or initiating a community solar project, the first step is to engage with local energy organizations, non-profits, and government agencies that can provide guidance, resources, and support. Researching existing community solar models and learning from their successes and challenges can offer valuable insights into project development and management.

Community solar projects and sharing not only offer a practical solution for expanding access to renewable energy but also embody the principles of cooperation, sustainability, and collective action towards a cleaner, greener future. By leveraging the power of the community, these projects can play a significant role in accelerating the adoption of solar energy, reducing greenhouse gas emissions, and building more resilient and sustainable communities.

CHAPTER 10

OTHER RESOURCES

Expanding your knowledge and capabilities in solar power goes beyond the hardware and technical specifications. A wealth of other resources is available to individuals interested in solar energy, offering support, information, and community engagement opportunities. These resources can help you navigate the complexities of solar power systems, stay informed about the latest advancements, and connect with like-minded individuals and organizations. Here's an overview of valuable resources for anyone involved in or considering solar power.

Educational Websites and Online Courses: Numerous websites and platforms offer in-depth articles, tutorials, and courses on solar energy. These can range from basic introductions to solar concepts to advanced technical training on system design and installation. Websites like the Solar Energy Industries Association (SEIA) and platforms like Coursera or edX host educational content developed by experts in the field.

Government and Non-Profit Organizations: Many government agencies and non-profit organizations provide comprehensive guides, toolkits, and case studies on solar power. The U.S. Department of Energy's Office of Energy Efficiency & Renewable Energy, for example, offers resources for understanding solar technologies, policy information, and funding opportunities. Non-profits like The Solar Foundation and international organizations such as the International Renewable Energy

Agency (IRENA) offer insights into solar trends, best practices, and research.

Local Solar Energy Associations and Clubs: Joining a local solar energy association or club can provide networking opportunities, access to workshops and seminars, and insights into local solar projects and initiatives. These groups often serve as a platform for sharing experiences, troubleshooting system issues, and advocating for renewable energy policies at the local level.

Solar Energy Consultants and Contractors: For those looking to install or upgrade a solar power system, consulting with a professional solar energy contractor can provide customized advice based on your specific needs and location. These experts can assess your property, recommend the appropriate system size and components, and navigate permitting and incentive applications.

Online Forums and Social Media Groups: Online communities offer a platform for asking questions, sharing experiences, and staying updated on solar energy news. Forums like Solar Panel Talk and social media groups dedicated to renewable energy can connect you with a global community of solar enthusiasts and professionals.

Mobile Apps: Various mobile apps are designed to help solar system owners monitor their system's performance, calculate potential solar energy production, and even create solar setups. Apps like PV Watts Calculator, Solar Panel Monitor, and others provide valuable tools at your fingertips.

Trade Shows and Conferences: Attending solar energy trade shows and conferences can provide exposure to the latest technologies, products, and services in the solar industry. These events offer the chance to meet

manufacturers, service providers, and other solar energy professionals face-to-face, gaining insights into emerging trends and innovations.
Leveraging these other resources can significantly enhance your understanding and engagement with solar power. Whether you're a homeowner exploring solar energy for the first time, a business looking to reduce energy costs, or an enthusiast eager to learn more, the wealth of available information and community support can guide you toward more informed decisions and successful solar energy projects.

Installation and Maintenance Checklists

Creating comprehensive checklists for the installation and maintenance of solar power systems ensures that every critical step is addressed, promoting safety, efficiency, and the long-term reliability of the system. These checklists serve as valuable tools for guiding both professional installers and homeowners through the critical phases of solar system setup and upkeep.

Installation Checklist

1. Site Assessment
- Evaluate roof structure and condition.
- Determine optimal placement for sunlight exposure.
- Check for potential shading issues.

2. System Design and Sizing
- Calculate energy needs based on historical consumption.
- Design system layout and component configuration.
- Select appropriate solar panels, inverter, and battery storage (if applicable).

3. Permitting and Documentation
- Obtain necessary local building and electrical permits.
- Ensure compliance with local codes and regulations.
- Prepare documentation for utility interconnection and incentives.

4. Component Procurement
- Purchase solar panels, inverter, mounting hardware, and other system components.
- Verify all components are delivered and in good condition.

5. Installation Process
- Securely mount solar panels and ensure proper orientation and tilt.
- Install the inverter and connect it to the solar panels and home electrical system.
- Set up battery storage and charge controller (for off-grid or hybrid systems).
- Implement electrical wiring, ensuring proper connections and safety measures.
- Ground the system components to protect against electrical surges.

6. System Testing and Commissioning
- Perform electrical tests to verify proper operation.
- Activate the system and conduct initial performance monitoring.
- Complete utility interconnection, if applicable.

Maintenance Checklist

1. Regular Inspections
- Conduct visual checks for damage or wear on solar panels, mounting hardware, and wiring.

- Inspect the inverter and battery storage for any signs of malfunction or degradation.

2. Cleaning Solar Panels:
- Schedule regular cleaning to remove dirt, debris, and other obstructions.
- Use appropriate cleaning methods to avoid damaging the panels.

3. System Performance Monitoring:
- Utilize monitoring software to track energy production and consumption.
- Compare performance against expected output and investigate any discrepancies.

4. Electrical System Checks:
- Verify all electrical connections are tight and free of corrosion.
- Test system components, such as the inverter and charge controller, for proper operation.

5. Battery Maintenance (if applicable):
- Check battery terminals for corrosion and ensure tight connections.
- Monitor battery charge levels and health, performing equalization charges as needed.

6. Vegetation Management:
- Trim trees or bushes that may cause shading on solar panels.
- Ensure clear access to sunlight for maximum energy production.

7. Record Keeping:
- Maintain detailed records of maintenance activities, system performance data, and any repairs or replacements.

- Keep warranties and product documentation accessible for reference.

Adhering to these installation and maintenance checklists can significantly enhance the efficiency, safety, and lifespan of a solar power system. Regular attention and care, guided by these comprehensive steps, ensure the system continues to provide clean, renewable energy while minimizing potential issues and downtime.

Solar Power System Troubleshooting: A comprehensive guide on diagnosing and fixing common solar system issues

Troubleshooting solar power systems requires a systematic approach to identify and resolve issues that may affect performance. Understanding common problems and their solutions can help maintain optimal operation and extend the lifespan of your solar setup. Here's a guide to diagnosing and fixing prevalent solar system issues.

Decreased Energy Production

Possible Causes: Dirt and debris on panels, shading, system aging, or component failure.
Solutions: Regularly clean solar panels to remove obstructions. Trim foliage to eliminate shading. Check system components for wear and tear, replacing any that are underperforming or faulty.

Inverter Issues

Possible Causes: Overheating, incorrect installation, electrical faults.

Solutions: Ensure adequate ventilation around the inverter to prevent overheating. Verify that all electrical connections are secure and correct.

Reset the inverter to clear any faults. If problems persist, consult the manufacturer or a professional.

Battery Storage Problems

Possible Causes: Overcharging, deep discharging, poor maintenance, age.
Solutions: Adjust charge controller settings to prevent overcharging and deep discharging. Perform regular battery maintenance, including checking connections and cleaning terminals. Replace batteries that are beyond their useful life.

Intermittent Power Supply

Possible Causes: Loose connections, damaged wiring, malfunctioning charge controller.

Solutions: Inspect all electrical connections and tighten any that are loose. Replace damaged wires. Check the charge controller for errors and reset or replace it if necessary.

No Power Production

Possible Causes: System not correctly connected, blown fuses or tripped breakers, complete component failure.

Solutions: Verify that the system is correctly connected to the power grid or home electrical system. Check for and replace any blown fuses or reset tripped circuit breakers. Inspect individual components (solar panels, inverter, charge controller) for failure and replace them as needed.

Monitoring System Alerts

Possible Causes: Component malfunction, system underperformance, or configuration errors.

Solutions: Review alerts and error codes via the system's monitoring interface. Consult the system's manual or technical support for specific error code troubleshooting and resolution steps.

Poor System Efficiency

Possible Causes: Suboptimal panel orientation, inverter inefficiency, outdated technology.

Solutions: Adjust the orientation of solar panels if possible to maximize sun exposure. Consider upgrading to a more efficient inverter. Evaluate the system for potential upgrades to newer, more efficient technology.

For complex issues or when in doubt, it's advisable to contact a professional solar technician. Many solar system components are under warranty, and professional diagnosis can prevent voiding these warranties. Additionally, working with electricity poses safety risks, and a professional can ensure repairs are conducted safely and effectively.

Maintaining a proactive approach to troubleshooting and regular maintenance can significantly enhance the performance and longevity of your solar power system, ensuring it continues to provide clean, **renewable** energy for years to come.

CHAPTER 11

ADVANCED SOLAR POWER TECHNIQUES

As solar power evolves, new techniques are being developed to enhance its effectiveness and broaden its usage. These innovative methods are more than just incremental upgrades; they revolutionize how energy is captured, converted, and stored. Examples such as hybrid systems that combine solar with other renewables, smart home integrations, and solar trackers highlight the progressive changes in the field. Such innovations are vital for optimizing solar panel performance and meeting the varied energy needs of today's consumers. Additionally, advanced troubleshooting and optimization further enhance the reliability and longevity of solar installations, increasing their practicality and appeal across diverse applications. Each development marks progress toward a more sustainable and efficient future, maximizing solar energy utilization.

Hybrid Systems: Using Solar, Wind, and Hydro Power Together

Combining solar, wind, and hydropower into hybrid systems offers a solution that optimizes the benefits of each power source while reducing its drawbacks, which is a significant advancement in renewable energy technology. This multi-source approach enhances the reliability of energy supply and reduces dependency on specific weather conditions, making it a robust and sustainable option for powering our homes, businesses, and communities.

The integration of solar, wind, and hydropower leverages the natural variations in the availability and intensity of these energy sources throughout the day and year. For instance, solar power is most abundant during sunny days, wind energy can be captured during windy conditions, which might occur during the day or night, and hydropower is consistent, depending on water availability, which can be predictable based on seasonal water flows. Since some of these sources may be powerful even in the case of a poor source, hybrid systems can offer a more steady and dependable power supply.

A key advantage of hybrid systems is their capability to deliver dependable power. Traditional single-source systems often face periods of downtime when the primary energy source is unavailable. For example, solar panels produce no power at night, and wind turbines are idle when there is no wind. However, a hybrid system can switch between sources or combine outputs, ensuring continuous energy production.

In regions where sunlight patterns are predictable, solar energy can be the primary energy source during the day, supplemented by wind or hydro in the absence of sunlight. Similarly, when cloud cover may reduce solar output during rainy seasons, hydropower can take the lead, supplemented by wind.

Minimizing reliance on specific weather conditions is another significant advantage. In traditional systems, a long spell of cloudy or windless days could disrupt the energy supply. Hybrid systems mitigate this risk by balancing the variabilities of each power source. For example, hydroelectric power typically has less day-to-day and seasonal variability and can be relied upon to produce energy when solar and wind conditions are not optimal. This not only ensures a stable power supply but also reduces the need for extensive battery storage, which can be expensive and have environmental impacts.

Real-world implementations of hybrid systems showcase their potential. For instance, in the Gorge Farm Energy Park in Naivasha, Kenya, a successful hybrid system integrates biogas, solar, and hydropower to produce 100% renewable energy. This farm uses solar panels during the day, hydropower generated by a nearby dam, and biogas produced from agricultural waste to provide a continuous power supply. Such an approach not only ensures constant energy production but also leverages local resources, reducing waste and enhancing sustainability.

Another example is the King Island Renewable Energy Integration Project in Tasmania, Australia, which combines wind, solar, and diesel with energy storage technologies. The system has successfully provided over 65% of the island's energy needs from renewable sources, with peaks reaching up to 100% at times. This initiative not only lowers greenhouse gas emissions and fuel consumption but also provides a model for comparable communities throughout the globe.

Moreover, the development of smart grids and advanced energy management systems further enhances the effectiveness of hybrid systems. These technologies allow for the intelligent distribution and utilization of energy based on real-time data and demand, optimizing the use of each power source and reducing waste. For instance, when excess power is produced, it can be diverted to recharge batteries or for other uses, such as heating water, thereby maximizing the efficiency of the energy produced.

Solar Power System Automation: Smart Home Integration

Integrating solar power systems with smart home technologies marks a significant leap toward enhancing home energy efficiency and management. This integration allows homeowners to harness the power of automation to optimize energy usage, control various aspects of their solar system remotely, and improve overall energy efficiency.

The core of solar power system automation lies in its seamless integration with smart home platforms, which enables homeowners to monitor and control their energy production and consumption through intuitive interfaces on their smartphones, tablets, or computers. This connectivity is typically achieved through IoT (Internet of Things) devices that communicate with each other over a home network.

Optimizing energy use is one of the main benefits of combining solar power systems with smart home technology. Smart home systems can automatically adjust the operation of appliances based on the availability of solar power, thereby maximizing the use of solar energy and reducing reliance on the grid.

For example, during peak solar production hours, smart home systems can prioritize running high-energy-consuming appliances like washing machines, dishwashers, and air conditioners. This not only utilizes solar energy efficiently but also lowers electricity bills by reducing grid energy consumption during peak tariff times.

Moreover, smart home integration allows for the remote management of solar power systems. Homeowners can monitor real-time data on energy production and consumption through their smart devices. They can receive instant notifications if the system underperforms or if there are issues requiring attention. This degree of monitoring makes sure that any possible issues can be resolved quickly, reducing downtime and optimizing the solar power system's performance.

Improving overall energy efficiency is another significant benefit of solar power system automation integrated with smart home technologies. Smart thermostats, for instance, can work in tandem with solar panels to regulate home heating and cooling efficiently. They can adjust the indoor temperature based on weather forecasts, the presence of people in the home, and the amount of solar energy being generated. This intelligent adjustment helps in maintaining optimal comfort while using less energy.

Popular smart home platforms that integrate with solar systems include Google Home, Amazon Alexa, and Apple HomeKit. These platforms offer extensive compatibility with various smart devices and provide users with the flexibility to set up automated routines. For instance, Google Home can integrate with solar inverters and battery systems to provide commands that maximize the use of solar energy based on predictive weather algorithms.

Amazon Alexa allows users to connect with devices like the Tesla Powerwall, enabling voice-activated control over solar storage systems. Homeowners can ask Alexa to switch from grid power to stored solar energy during a power outage or when grid prices are high, ensuring energy cost savings and continuous power supply.

Apple HomeKit's integration with solar power systems extends to automating window shades and lights based on the amount of sunlight penetrating the home. This not only helps in using natural light effectively but also in preserving the energy stored in solar panels.

In addition to these, platforms like Samsung SmartThings and IFTTT (If This Then That) offer further integrations, connecting various sensors and switches to the solar power setup, which can automate lights, HVAC systems, and even garden irrigation based on the energy available from the solar panels.

As smart home technologies advance, they promise to further revolutionize the way we manage and consume energy in our homes, bringing us closer to a more sustainable and energy-efficient future. Homeowners may adopt a proactive approach to energy management and make sure that renewable energy sources are exploited to their maximum potential by combining the conveniences of contemporary technology with solar power systems.

Maximizing Efficiency with Solar Trackers

Solar trackers represent a dynamic technology designed to enhance the energy efficiency of photovoltaic (P.V.) systems by enabling solar panels to follow the sun's trajectory across the sky. This continual alignment with the sun optimizes the angle of incidence between the solar panels and sunlight, maximizing the amount of solar radiation captured and converted into electricity.

Solar trackers are primarily used to move solar panels around during the day and, in certain situations, over the seasons. As the sun moves, the trackers tilt and rotate the panels to maintain the best possible angle to capture sunlight. This optimizes the panels' exposure to direct sunlight, significantly increasing their efficiency compared to stationary setups.

Stationary solar panels, typically mounted at a fixed angle, are oriented based on average sunlight conditions and do not adjust to changes in the sun's position. While this setup benefits from simplicity and lower upfront costs, it does not capture the maximum possible energy. In contrast, solar panels equipped with trackers can produce up to 10-25% more electricity depending on the geographic location and specific environmental conditions. The generation of energy might be significantly impacted by this efficiency gain, particularly in areas with abundant solar radiation.

There are several types of solar trackers, each suited to different scenarios and needs. The two main categories are single-axis trackers and dual-axis trackers. Single-axis trackers rotate on one axis, moving either in an east-west direction or a north-south direction. This type is most effective in regions closer to the equator where the sun's path remains relatively consistent throughout the year.

Dual-axis trackers, on the other hand, provide rotation both horizontally and vertically, allowing for precise alignment with the sun as it changes its path across different seasons. This type is ideal for higher latitudes where the sun's elevation changes significantly with the seasons. Within these categories, there are specific types of trackers.

Horizontal single-axis trackers are common in large-scale solar farms where rows of solar panels tilt together in unison to follow the sun's movement from east to west. Vertical single-axis trackers, which are less common, align the panels from north to south and are better suited for higher latitudes.

Tilted single-axis trackers and polar-aligned trackers offer more flexibility in terms of angling the panels toward the sun. For the most advanced tracking, dual-axis trackers adjust both azimuth and elevation, providing the optimal angle at all times, which maximizes energy capture and efficiency.

The ideal scenarios for using solar trackers vary based on geographical location, solar goals, and economic considerations. In commercial solar projects or utility-scale installations, the increased energy output from using trackers can justify the higher initial investment and maintenance costs due to the significant boost in power generation efficiency. For residential applications, the decision to use solar trackers depends on factors like roof space, budget, and the homeowner's energy requirements.

While the increased cost and complexity of installing solar trackers might not be justified in every residential scenario, they can be highly beneficial in off-grid systems where maximizing energy input is critical. Furthermore, the integration of solar trackers in regions with high solar irradiance can dramatically enhance the performance of solar installations. In these environments, the extra energy produced can provide a quicker return on investment, making trackers an attractive option despite the higher initial costs. Conversely, in areas with lower solar irradiance or where space constraints limit the feasibility of installing trackers, stationary panels might be a more cost-effective solution.

Integrating Solar Power with Electric Vehicle Charging

Integrating solar power systems with electric vehicle (E.V.) charging stations presents an innovative and sustainable solution to meet the growing demand for green transportation options. Eco-friendly charging options are becoming more and more necessary as the use of electric vehicles increases. Utilizing solar energy to power E.V.s harnesses a renewable resource, significantly reducing the carbon footprint associated with vehicle charging.

Installing solar photovoltaic (P.V.) panels to collect solar energy and transform it into electricity is the first step in integrating solar power with electric vehicle (E.V.) charging. This system can be implemented in various settings, including residential, commercial, and public charging stations. The basic setup includes solar panels, an inverter to convert the generated D.C. electricity into A.C. electricity, battery storage to hold excess power, and a charging station to deliver the power to the vehicles.

One of the primary benefits of using solar power for E.V. charging is the significant reduction in greenhouse gas emissions. Traditional E.V. charging often relies on electricity generated from fossil fuels, which undermines the environmental benefits of electric vehicles. By contrast, solar-powered charging stations use clean energy, ensuring that the entire lifecycle of E.V. operation remains environmentally friendly.

Another advantage is cost efficiency. Although the initial setup for solar charging stations can be higher due to the cost of solar panels and storage batteries, the ongoing costs are significantly lower. After installation, solar panels generate almost free energy for a considerable amount of time—usually 25 to 30 years. For homeowners with E.V.s, this means the cost of charging their vehicle can be significantly reduced, delivering savings on electricity bills.

In addition, the combination of solar energy with E.V. charging promotes energy independence. People and companies that generate their own electricity can become less dependent on the grid, which is especially useful in areas with high electricity prices or during peak hours. Additionally, during power outages or in remote areas without reliable access to grid electricity, solar-powered E.V. charging stations can provide an uninterrupted power supply.

In public spaces and workplaces, solar-powered E.V. charging stations also serve as a visible commitment to sustainability, enhancing the corporate image and potentially attracting customers or employees who value environmental responsibility. Furthermore, these installations can be equipped with smart charging technology, which optimizes charging schedules based on energy production and demand patterns, further enhancing efficiency and reducing costs.

Looking into the future, several trends are likely to shape the development of solar-powered E.V. charging. The incorporation of intelligent grid technology, which enables the regulation of energy flow based on real-time data, is one notable trend. This technology can manage the distribution of solar energy not only to charge E.V.s but also to supply residential and commercial buildings depending on current demand, thereby maximizing the use of generated solar power.

Another trend is the development of portable solar chargers, which can be deployed quickly and easily in various locations, providing great flexibility for E.V. charging. These portable units can be particularly useful for temporary setups at events or in rental properties, offering a versatile and eco-friendly charging solution.

Furthermore, improvements in battery technology are anticipated to improve solar energy storage, resulting in more dependable and efficient solar-powered E.V. charging stations. Innovations such as solid-state batteries could offer higher energy density and faster charging times, which would significantly improve the practicality and appeal of solar-powered E.V. charging.

As the world moves towards a more sustainable future, the role of solar energy in powering electric vehicles is set to grow. The integration of solar power with E.V. charging not only supports environmental goals but also offers economic and practical benefits, positioning it as a critical component of future transport infrastructure. With ongoing technological advancements and more environmental awareness, it is expected that the number of solar-powered electric car charging stations will increase, taking us one step closer to an emission-free future.

Advanced Troubleshooting and System Optimization

Troubleshooting and optimizing solar power systems are essential for maintaining efficiency and extending operational life. Effective troubleshooting, regular maintenance, and strategic optimization can significantly enhance a solar system's performance.

When troubleshooting common issues in solar power systems, several key challenges need addressing. Reduced power output is a frequent concern, often caused by obstructions like dirt, debris, or snow that block sunlight from reaching the solar cells. To address this, solar panels should be cleaned regularly to remove any materials that could impair their function.

Additionally, it's important to periodically check for new shading, such as from growing trees or new constructions, which can also reduce efficiency. Examining the panels for physical damage, such as scratches or fractures, is essential in addition to cleaning and managing the shade, as these flaws might affect the panels' effectiveness.

Inverter failures present another critical area of concern. By transforming the D.C. power produced by the solar panels into A.C. electricity, the inverter plays a crucial part. Signs of inverter problems include unusual noises, error messages, or a complete stop in energy production. Initially, resetting the inverter may resolve some issues; however, if problems persist, checking electrical connections or consulting a professional may be necessary. For systems that include battery storage, maintaining the health of these batteries is imperative.

Symptoms of battery issues include the inability to hold a charge or rapid discharge. Keeping battery terminals clean and ensuring they're tightly connected helps, as does making sure the batteries operate within recommended temperature ranges to avoid performance degradation.

Regular maintenance is crucial for the system's longevity and efficiency. This includes cleaning the panels to ensure they receive maximum sunlight and conducting annual inspections to check all hardware and electrical connections. These inspections help identify and mitigate potential failures early on.

Optimizing a solar power system involves several strategies to boost efficiency and lifespan. Upgrading to newer inverter technology can offer better efficiency and features like enhanced monitoring, which aids in managing the system more effectively. Implementing maximum power point tracking (MPPT) controllers is another strategy that improves energy conversion efficiency. MPPT controllers adjust the load continuously to maximize power output under varying sunlight conditions.

Battery storage technology also plays a critical role in optimization. Upgrading to advanced battery technologies such as lithium-ion can provide better performance, especially in variable temperature conditions, and offer longer lifespans than traditional battery types. Additionally, managing the thermal environment of solar panels and batteries through passive or active cooling systems can prevent overheating and enhance overall system performance, which is particularly useful in hotter climates where high temperatures can reduce efficiency.

By integrating these approaches—thorough troubleshooting, diligent maintenance, and strategic optimizations—owners of solar power systems can ensure that their installations perform optimally. This not only maximizes the utility and return on investment of their solar systems but also supports the broader adoption and effectiveness of renewable energy solutions. Such practices not only sustain the technical health of the system but also contribute to the environmental benefits of utilizing renewable energy sources.

CHAPTER 12

FINANCIAL AND ENVIRONMENTAL IMPACT ANALYSIS

Exploring the financial and environmental impact of solar power systems unveils a dual advantage that extends beyond mere energy generation. This analysis delves into the long-term economic benefits, revealing how solar investments can significantly reduce utility bills and provide a stable, low-cost energy supply for decades. Concurrently, the environmental benefits are profound. As a clean and sustainable resource, solar energy is essential for cutting carbon emissions and limiting the environmental impact of using conventional fossil fuels. By evaluating both the financial savings and the positive environmental effects, this examination provides a comprehensive understanding of the multifaceted value that solar power brings to individuals, communities, and the planet.

Long-Term Financial Benefits of Solar Power

Investing in solar power is a strategic decision that not only embraces renewable energy but also yields substantial financial benefits over time, with initial costs offset by ongoing utility bill reductions, inflation protection, and various government incentives that improve the overall return on investment.

Installing solar panels has a significant financial advantage in the form of lower power costs. Once a solar power system is operational, it generates free electricity for the property's use, directly decreasing the energy needed from the grid. Depending on the size of the solar system and the patterns of energy use, this can save monthly power costs for many homes and companies by 50% to 100%. As utility prices grow over time, the savings may increase significantly. Solar panels allow homeowners and businesses to lock in energy costs at a lower rate, providing a hedge against future price increases driven by inflation and fluctuating market conditions.

Inflation plays a critical role in the financial dynamics of energy costs. Traditionally, the price of electricity has increased at a rate that outpaces general inflation due to rising fuel costs, aging infrastructure, and increased regulation. Investing in solar power freezes a portion of their electricity rate, thereby insulating themselves from these increases. As a result, the savings grow each year as the difference between the fixed cost of solar-generated electricity and the escalating market price of grid electricity widens. This aspect of solar investment provides immediate financial relief and long-term economic security.

The return on investment (ROI) for solar power systems is another crucial aspect, demonstrating the economic soundness of solar panels as a capital investment. The average payback period for a solar installation can vary widely, typically between 5 to 10 years, depending on the initial cost, the amount of sunlight the location receives, and local electricity rates.

After breaking even, the solar system continues to generate free electricity, which translates into pure savings, thus enhancing the property's value both economically and environmentally. Moreover, solar panels have a long lifespan, usually 25 to 30 years, which means they can provide two to three decades of electricity at a fraction of the cost of buying from the power grid.

Government incentives and rebates play a pivotal role in enhancing the financial benefits of solar power. Many governments worldwide offer substantial incentives to encourage the adoption of solar energy. These can include direct rebates that reduce the upfront cost of solar systems, tax incentives such as tax credits and accelerated depreciation, and other financial benefits like net metering and feed-in tariffs. Solar system owners can further offset the cost of any power they must purchase during times of poor solar production by using net metering to sell excess electricity back to the grid at retail rates. Similarly, feed-in tariffs provide long-term contracts to solar system owners at guaranteed rates for the electricity they produce, which can significantly improve the financial return.

All these factors—savings on electricity bills, protection against inflation, return on investment, and government incentives—converge to make solar power a wise financial decision. Beyond the environmental benefits, the financial rationale for investing in solar technology is compelling. As energy prices continue to rise and the cost of solar installations decreases, the economic argument for solar power becomes even more persuasive.

This shift not only reflects the growing accessibility and affordability of solar technology but also aligns with broader economic trends toward sustainability and energy independence. Through detailed analysis and practical examples, the long-term financial advantages of solar power are clear, offering not just a pathway to a greener planet but also a sound investment strategy that yields considerable economic returns over time.

Environmental Impact and Sustainability

Using solar power is a critical step in the global fight against climate change because it has a smaller environmental impact than energy sources that rely on fossil fuels. The environmental benefits of solar power stretch from large reductions in greenhouse gas emissions to the protection of natural resources, making it a cornerstone of sustainable energy initiatives.

The main benefit of solar power for the environment is that it produces clean, renewable energy without emitting any carbon dioxide (CO_2).

Solar power plants generate energy without burning fossil fuels, preventing the direct emissions of dangerous pollutants and greenhouse gases. This is in contrast to traditional power plants that burn coal, natural gas, or oil. This is extremely important since burning fossil fuels contributes significantly to both air pollution and climate change. Utilizing solar energy, solar power offers a workable substitute that lessens these negative effects on the environment.

Studies estimate that an average residential solar panel system can reduce carbon emissions by approximately three to four tons annually. Over the typical 30-year lifespan of a solar installation, this translates to preventing around 90 to 120 tons of CO2 from entering the atmosphere.

When scaled up to include commercial and utility-scale solar operations, the potential for carbon reduction is massive, offering a clear path toward meeting global carbon reduction targets. Moreover, solar energy helps conserve finite natural resources that would otherwise be used for energy production. Traditional electricity generation is heavily reliant on water, particularly for cooling thermal power plants, which places a significant strain on this vital resource.

In contrast, photovoltaic cells used in solar panels require no water to generate electricity, making solar power a precious technology in regions facing water scarcity. This aspect of solar energy not only aids in water conservation but also reduces the water pollution typically associated with conventional power generation.

Although there are environmental concerns associated with the manufacture and disposal of solar panels, industry trends indicate that more sustainable methods will prevail. More recyclable components and less hazardous materials are being used as a result of advancements in solar technology. Companies are now actively recycling old solar panels and repurposing the silicon and precious metals they contain, further reducing the environmental footprint of solar power systems.

Another place where solar energy's advantages are evident is in the way it affects ecosystems and wildlife. Solar farms generally have a much lower impact on their surrounding environments than large-scale hydroelectric projects or fossil fuel operations that can devastate local wildlife habitats. With thoughtful planning, solar installations can be developed on degraded or contaminated lands unsuitable for agriculture or habitation, minimizing their ecological disruption.

Some solar farms have even been designed to double as pollinator-friendly environments, supporting bee and butterfly populations that are crucial to the health of local ecosystems. Finally, integrating solar power into urban environments showcases its role in sustainable urban development. Rooftop solar installations and solar-powered street lighting reduce urban heat island effects and lower the overall energy demand of buildings, contributing to more sustainable cityscapes.

Case Studies: Successful Off-Grid Solar Installations

Successful off-grid solar installations offer insightful case studies into the adaptability and efficacy of solar power across diverse environments and needs. From remote rural homes to bustling commercial hubs, the implementation of off-grid systems underlines a significant shift towards sustainable energy independence. Each case study below provides a detailed look at how various challenges were addressed, demonstrating the robust potential of solar technology.

1.Residential Off-Grid Solar Installation in the Scottish Highlands

In the rugged terrain of the Scottish Highlands, a residential property sought to establish energy independence due to the high cost and impracticality of connecting to the national grid. The homeowners opted for a comprehensive off-grid solar system supplemented by a wind turbine, considering the region's limited sunlight during winter months.

Challenges: The primary challenge was the area's harsh weather conditions, which often included heavy cloud cover and strong winds. Additionally, the remote location meant that all components needed to be highly reliable, and maintenance had to be minimal.

Solutions: The system combined a 5 kW solar P.V. array with a 3 kW wind turbine, ensuring energy availability during sunny and windy conditions. A bank of deep-cycle batteries stored excess energy and a backup biodiesel generator was installed to provide power during prolonged poor weather conditions. The hybrid system allowed for a continuous power supply, with the solar panels providing sufficient energy during the summer and the wind turbine taking advantage of the higher winter winds.

2. Commercial Off-Grid Solar Project in the Australian Outback

A mining company operating in the Australian Outback implemented an off-grid solar system to power their exploration site, reducing reliance on diesel generators, which were expensive due to fuel transport costs.

Challenges: The extreme temperatures and isolated location presented significant challenges. Solar installations needed to be robust enough to withstand intense heat and dust without frequent maintenance.

Solutions: The installation featured high-efficiency solar panels specially designed to perform in high-temperature climates, paired with a state-of-the-art cooling system to prevent overheating. The system included a large-scale battery storage unit to manage energy supply reliably, even during periods of low sunlight. By reducing generator use, the company significantly cut fuel costs and minimized its environmental footprint.

3. Rural Off-Grid Solar System in the Kenyan Savannah

In rural Kenya, a community-based project saw the installation of an off-grid solar system to power a local health clinic and school, providing reliable electricity to crucial community services for the first time.

Challenges: The main challenges were the lack of local infrastructure for energy, the need for a system that local technicians could easily maintain, and the need for training residents to manage the system.

Solutions: The project utilized simple, durable solar technology with easy-to-replace components. Training programs were established for residents, focusing on system maintenance and basic troubleshooting. The installation provided essential electrical services and facilitated educational opportunities by powering computers and other educational equipment.

4. Off-Grid Solar Powered Eco-Resort in the Brazilian Amazon

An eco-resort in the Brazilian Amazon embraced off-grid solar to minimize its environmental impact and align with its sustainability goals. The resort was located far from grid access, surrounded by dense rainforest.

Challenges: High humidity and frequent rain risk solar panel efficiency and system components. The solution needed to blend with the environmental and aesthetic values of the eco-resort.

Solutions: The installation included elevated solar panels above the canopy level to gain maximum sunlight exposure. The system was integrated with a rainwater collection system that also helped to keep the solar panels clean and efficient. A series of batteries stored excess solar power, ensuring the resort had sufficient energy even during overcast days.

5. Off-Grid Solar Installation in a Mountainous Ski Lodge in Colorado

A ski lodge in Colorado opted for an off-grid solar solution to power its operations, including lifts and lighting, reducing its environmental impact and operating costs.

Challenges: The primary challenge was dealing with heavy snowfall, which could cover panels and drastically reduce their effectiveness. The remote mountain location also meant that system components had to be exceptionally reliable.

Solutions: The lodge used a solar tracking system to adjust the panels' angle for maximum sunlight capture throughout the day. Vertically oriented solar panels were installed to help shed snow more effectively. For added power resilience, a small hydroelectric generator powered by a nearby stream and battery storage were used.

These case studies demonstrate the versatility and adaptability of off-grid solar systems in providing reliable, sustainable energy solutions across various environments and applications. Each installation highlights the importance of tailored approaches considering local conditions and needs, ensuring that solar power remains a crucial player in the global transition to renewable energy.

Calculating ROI for Solar Investments

There are several financial considerations involved in calculating the return on investment (ROI) of solar power systems, ranging from upfront expenditures to ongoing savings and incentives. This systematic approach provides potential investors with a clear picture of the economic value of their investment in solar energy. Understanding ROI is crucial as it helps determine the feasibility and profitability of a solar installation over its operational lifespan.

The first step in calculating a solar power system's ROI is compiling all initial costs. These costs include the purchase price of all solar panels, inverters, mounting equipment, and wiring. It is also necessary to account for the installation expenses, which might differ significantly based on the system's complexity, location, and labor prices. Gathering accurate cost data to ensure the ROI calculation reflects the actual financial investment required is important.

Next, consider the maintenance costs associated with the solar power system. While solar panels are renowned for their low maintenance requirements, occasional cleaning and periodic checks are necessary to ensure optimal performance. In some cases, inverters may need to be replaced after 10 to 15 years, which should be included as a future cost. Estimating these costs upfront provides a more comprehensive view of the ongoing financial commitment to operating a solar power system.

Energy savings constitute the most significant variable in the ROI calculation and are influenced by several factors. To estimate these savings, calculate the average home or business electricity usage and then determine how much of that demand can be met by the solar system. Using local electricity rates, project the cost savings over each month. These calculations must consider potential changes in energy rates over time, often assuming a conservative annual increase in utility rates to reflect historical trends.

Another crucial element in the ROI calculation is the benefit derived from any available incentive programs. Many governments offer solar energy installation tax incentives, rebates, or credits. These can significantly reduce the initial outlay and improve the overall ROI. For example, a federal tax credit may cover a percentage of the total cost of the solar system, including installation. Additionally, some municipal governments and utilities provide solar customers with incentives like cash back, exemptions from property taxes, or lower electricity prices. It is necessary to quantify each of these incentives and incorporate them into the ROI analysis.

The next stage is to figure out the payback period, or the amount of time it will take for the savings from the solar system to equal the initial investment, once all expenses and savings have been determined. This is done by dividing the total initial costs by the annual financial benefits (annual energy savings plus any incentives). The result is the number of years it will take for the investment to pay for itself.

Once the payback period has been established, calculating the overall return on investment involves a more extended projection. Typically, solar panels have a warranty of 25 years, but they can continue to function beyond this period, albeit at reduced efficiency. To determine the ROI, project the total savings over the desired time frame, subtract the combined initial and maintenance costs, and divide this net benefit by the initial costs. This figure, expressed as a percentage, represents the ROI and clearly indicates the financial returns expected from the solar investment.

Additionally, it is prudent to consider the increased property value resulting from the solar installation. Research indicates that houses equipped with solar panels get a higher price on the market than those without. This increase in property value should also be factored into the overall ROI, as it represents an additional financial benefit that can significantly enhance the attractiveness of the investment.

Calculating the ROI of a solar power system requires a detailed analysis of costs, savings, and incentives. By methodically examining these factors, investors can make informed decisions about solar investments, aligning financial goals with environmental stewardship. The comprehensive calculation highlights the direct financial returns and underscores the broader economic and ecological benefits of transitioning to solar energy.

Understanding Carbon Footprint Reduction

Understanding how solar power systems contribute to carbon footprint reduction involves examining the relationship between energy generation, greenhouse gas emissions, and the shift from fossil fuels to renewable energy sources. Solar power plays a pivotal role in mitigating climate change by providing a clean alternative to traditional energy sources, which are significant contributors to global carbon emissions.

Photovoltaic (P.V.) cells are used in solar power systems to convert sunlight into electrical power. This process is inherently clean, as it involves no burning of fossil fuels, no emissions of greenhouse gases during operation, and minimal environmental disruption. The primary environmental benefit of deploying solar panels is the direct reduction in the need for electricity generated from coal, natural gas, or oil. By offsetting energy production from these carbon-intensive sources, solar power effectively reduces the overall emissions of carbon dioxide (CO_2), methane (CH_4), and other pollutants that contribute to global warming and air pollution.

To quantify the impact of solar power on carbon footprint reduction, consider a typical residential or commercial solar installation. For instance, the average American home consumes approximately 10,972 kilowatt-hours (kWh) of electricity annually. If we assume this electricity is sourced from a conventional grid mix heavily dependent on fossil fuels, significant emissions are associated with this energy use. By contrast, a residential solar system capable of generating an equivalent amount of electricity can dramatically decrease these emissions.

In terms of specific numbers, a typical residential solar system size ranges from 5 to 10 kilowatts (kW) and produces around 6,000 to 12,000 kWh of electricity annually, depending on geographic location, panel efficiency, and other factors. The Environmental Protection Agency (EPA) provides a greenhouse gas equivalencies calculator, which estimates that every kWh of electricity generated from non-renewable sources in the U.S. results in approximately 0.92 pounds of CO_2 emitted into the atmosphere. Thus, by developing, for example, 10,000 kWh per year, a solar system can prevent around 9,200 pounds (or 4.6 tons) of CO_2 from being emitted annually.

Commercial solar installations, typically larger, can have an even more substantial impact. These systems may range from tens to hundreds of kilowatts, with higher electricity production. A medium-scale commercial solar system generating around 100,000 kWh annually could avoid approximately 92,000 pounds (or 46 tons) of CO_2 emissions annually. This reduction is equivalent to the CO_2 emissions from consuming over 4,700 gallons of gasoline or burning over 45,000 pounds of coal.

Beyond the direct reduction of emissions, solar power systems contribute to long-term sustainability and environmental protection. By reducing dependence on finite natural resources and decreasing air and water pollution associated with fossil fuel extraction and combustion, solar energy helps preserve ecosystems and improve public health. Moreover, the increase in solar energy capacity helps drive technological advancements and cost reductions, making sustainable energy more accessible and practical.

Deploying solar power systems across residential, commercial, and industrial sectors is a critical strategy in global efforts to reduce carbon footprints. As these systems replace or reduce reliance on fossil fuels, they decrease the amount of greenhouse gases released into the atmosphere, thus playing a crucial role in combating climate change. With continued advancements and increased adoption, the role of solar energy in achieving carbon neutrality and fostering a sustainable future becomes increasingly significant, highlighting the essential nature of renewable energy solutions in the global environmental strategy.

CHAPTER 13

SOLAR POWER FOR SPECIALIZED APPLICATIONS

Solar power's versatility extends beyond typical residential and commercial applications, adapting seamlessly to more specialized settings. This adaptability allows for the effective use of solar energy in environments ranging from compact urban homes to expansive agricultural fields. In each scenario, solar power brings unique benefits, catering to different users' specific energy needs and constraints. Whether it's providing reliable power in remote locations, supporting sustainable farming practices, or ensuring energy independence for mobile units, the potential applications of solar technology are vast. By exploring these diverse applications, we can better understand how solar energy can be tailored to meet a wide range of practical requirements, further cementing its role as a pivotal solution in the global shift towards sustainable energy.

Solar Power for Tiny Homes and Portable Units

Solar energy is increasingly integral to powering tiny homes and portable units, aligning perfectly with the modern drive towards minimalism, mobility, and environmental sustainability. The distinct appeal of solar power for such applications lies in its ability to provide a reliable source of energy that supports the ethos of independence and flexibility characteristic of these living spaces.

Design Considerations for Tiny Homes

Tiny homes, part of a movement embracing less than 500 square feet of space, require energy solutions that are both efficient in performance and space. Designing a solar power system for such limited areas demands high-efficiency photovoltaic panels that provide more power output per square foot than traditional panels. These are ideal as they maximize the limited roof area available in tiny homes.

Moreover, the architectural variety of tiny homes—from those built on foundations to those constructed on trailers—requires flexible installation options. Solar technology has adapted to this need with innovations such as thin-film solar panels and solar shingles. Thin-film panels, known for their lightweight and flexible, can be installed on surfaces unsuitable for traditional panels. Solar shingles, on the other hand, offer a dual purpose by acting as both a roofing material and a solar energy generator, maintaining the aesthetic appeal of the tiny home without compromising its functionality.

Energy Efficiency and Storage

Energy efficiency is crucial in tiny homes, where space is at a premium and every watt counts. Solar systems in these settings are often paired with energy-efficient appliances to minimize energy consumption. LED lighting, energy-efficient refrigerators, and propane or solar water heaters are commonly used to reduce the overall energy load.

Integrating battery storage is also essential for ensuring a continuous power supply, especially in off-grid setups. Modern solar systems for tiny homes typically include a set of batteries that store excess electricity generated during the day. These batteries are becoming more compact, efficient, and capable of providing power throughout the night and during overcast days, enhancing the home's energy independence.

Portable Solar Power Units

For portable units, which include mobile offices, emergency power supplies, and camping setups, the design of solar systems focuses on portability and ease of use. These systems feature lightweight, durable components that withstand frequent movement and varied environmental conditions.

Portable solar units are often designed with modularity, allowing users to scale their power supply based on immediate needs. This modularity is facilitated by connectors that enable additional solar panels and battery packs to be easily integrated as needed. Moreover, many portable solar systems come equipped with controllers that manage power input and output, ensuring devices are charged safely and efficiently.

Mobility and Flexibility Benefits

The mobility provided by solar power is particularly beneficial for individuals who frequently relocate or live a nomadic lifestyle. For instance, solar-powered tiny homes on wheels can be moved from one location to another without losing access to power. Similarly, portable solar units can be used in various outdoor settings, providing reliable power without fuel-based generators, which are noisy and pollute the environment.

Environmental Impact

Adopting solar power in tiny homes and portable units significantly reduces the carbon footprint of these dwellings. Solar energy does not emit pollutants or greenhouse gases, making it a clean alternative to conventional energy sources. This is particularly appealing to those who choose tiny homes and portable units as a way to reduce their environmental impact and promote sustainability.

Challenges and Future Prospects

Despite the many benefits, there are challenges, such as the initial cost of solar systems and the need for sunlight exposure, which can limit usability in heavily shaded or cloudy areas. However, advances in photovoltaic technology and the decreasing cost of solar components make solar more accessible and efficient.

Looking ahead, the integration of solar power in tiny homes and portable units is expected to grow, driven by technological advancements and increasing environmental consciousness. Innovations like bi-directional charging, which allows electric vehicles to be charged from home solar systems and vice versa, and the integration of smart home technologies for energy management are likely to enhance the functionality and appeal of solar-powered living spaces.

Using Solar Power in Agricultural Settings

Solar energy is increasingly becoming integral to modern agriculture, offering innovative solutions that enhance efficiency while promoting sustainability. As farms and agricultural operations seek ways to reduce costs and minimize their environmental impact, solar power is a versatile and eco-friendly option. By harnessing the sun's energy, farms can power a wide range of operations, from irrigation systems to the daily electrical needs of barns and outbuildings, all while contributing to a more sustainable agricultural model.

One of the most common applications of solar energy in agriculture is in the operation of water pumps. Traditional water pumping systems are often powered by diesel generators, which are not only costly but also contribute to significant CO_2 emissions. Solar-powered water pumps offer a clean, reliable, and cost-effective alternative. These systems are particularly advantageous in remote or rural areas with limited access to electricity. Solar pumps can operate anywhere there is sunlight, providing farmers with a consistent water supply for irrigation, livestock, and other needs without the ongoing fuel costs and maintenance associated with diesel pumps.

In addition to water pumps, solar panels are used to power a variety of other farm operations. This includes providing electricity for lighting barns, outbuildings, and greenhouses, which is essential for extending working hours and enhancing security on the farm. Solar energy can also power ventilation systems in animal enclosures, helping to maintain a healthy environment by regulating temperature and humidity levels. Furthermore, solar panels can supply energy for electric fencing and security cameras, adding extra safety and efficiency to farm management.

The financial benefits of using solar power in agricultural settings are significant. The long-term savings on energy bills offset the initial setup cost of solar installations. Once installed, solar panels require minimal maintenance and provide a reliable power supply that is not subject to fluctuations in fuel prices. This stability is precious in agricultural operations, where budgeting and cost management are critical to profitability. Moreover, in many regions, government incentives such as grants, tax breaks, and rebates are available to help offset the initial costs of solar power systems, making solar energy an even more attractive option for farmers.

Solar power not only reduces operational costs but also promotes sustainable farming practices. By decreasing reliance on fossil fuels, farms can significantly lower their carbon footprint, contributing to environmental conservation. Solar energy use in agriculture aligns with global efforts to combat climate change, enhancing a farm's marketability to eco-conscious consumers and helping farmers meet regulatory standards concerning environmental impact.

Furthermore, the use of solar energy in agriculture supports energy independence. Farms that generate their power are less vulnerable to disruptions in the energy grid and can operate more autonomously. This resilience is crucial in facing challenges such as power outages or increases in energy costs.

Overall, integrating solar power into agricultural settings offers many benefits, from cost savings and enhanced energy independence to contributions toward environmental sustainability. As technology advances and becomes more accessible, an even greater number of agricultural operations will likely turn to solar energy, recognizing it as a critical component of modern, efficient, and sustainable farming.

Solar Water Pumping Systems

Solar water pumping systems represent a significant advancement in sustainable agriculture and remote water management, utilizing the sun's energy to operate pumps that can transfer water over vast distances without relying on conventional power grids. These systems are particularly beneficial in remote areas with limited or non-existent access to reliable electricity. Understanding the components and operation of solar water pumping systems and their inherent advantages can help individuals and communities make informed decisions about their water management strategies.

The solar panel array is at the core of a solar water pumping system, which captures solar energy and converts it into electrical power. This array is connected to an electric motor that drives a water pump, typically a submersible or surface pump, depending on the application. The system often includes a controller to optimize the pump's operation, protecting it from running dry and managing the power supply to maximize efficiency. Additionally, many systems are equipped with batteries to store excess solar energy, ensuring water can be pumped even during overcast conditions or at night. However, many modern systems prefer direct solar operation without batteries to reduce costs and maintenance.

The operation of solar water pumping systems is straightforward yet highly efficient. During peak sunlight, the solar panels generate electricity that powers the pump. As sunlight varies, the system adjusts the pumping rate. This direct relationship between sunlight and water pumping capacity naturally aligns water supply with the daily cycle, which can be particularly advantageous for agricultural applications like irrigation during the warmer parts of the day.

One of the primary advantages of solar water pumping systems over traditional systems is their independence from the power grid and fossil fuels. This not only reduces operational costs—eliminating expenses for diesel fuel and electricity—but also enhances reliability in remote locations. Additionally, solar water pumps emit no pollutants, offering an environmentally friendly alternative to conventional diesel pumps, which produce carbon emissions and noise pollution.

Another advantage is the low maintenance requirement. Solar water pumps have fewer moving parts than diesel-powered pumps and do not require regular fuel or extensive mechanical maintenance. This reliability and ease of use make solar pumps a practical choice for remote communities and farms.

When selecting a solar water pumping system, several factors need to be considered to ensure the setup meets specific water needs and adapts to geographical conditions. The first step is to assess the daily water requirements, including water for irrigation, livestock, or domestic use. This assessment helps determine the pump size and the solar panel array capacity needed to meet these needs.

Geographical conditions also play a crucial role in system selection. In areas with high solar insolation, smaller solar arrays can be effective. In contrast, regions with lower sunlight levels might require larger arrays or additional panels to achieve the same pumping capacity. The depth of the water source is another critical factor. Deeper water sources require pumps with higher lift capabilities, which can influence the type of pump and the size of the motor.

For those looking to implement a solar water pumping system, consulting with professionals who can perform an on-site evaluation is advisable. These experts can offer specific advice on the optimal system configuration, considering the local climate, water table depth, and seasonal variations in sunlight and water demand.

Overall, solar water pumping systems provide a robust solution for managing water resources in various settings. Their cost-effectiveness, environmental benefits, and compatibility with remote operations make them a compelling choice for modern water management needs. As technology advances and the cost of solar components continues to decrease, these systems will likely become even more accessible and popular, representing a sustainable step forward in water resource management.

Solar Energy for Emergency Preparedness

Solar power has become a cornerstone in emergency preparedness strategies across the globe, offering a reliable and renewable energy source during unforeseen disasters or crises. The role of solar energy in these scenarios is critical, as it provides essential power when natural disasters, infrastructure failures, or other emergency situations compromise traditional electricity grids.

The independence of solar power systems from conventional grid infrastructure is one of their primary advantages in emergencies. These systems can operate autonomously, ensuring essential services such as lighting, communication, and medical equipment can continue functioning even when the grid is down. This feature is particularly crucial during extended power outages, which are common after major disasters like hurricanes, earthquakes, or floods.

A typical solar energy system for emergency preparedness includes several key components: photovoltaic (P.V.) panels, batteries for energy storage, an inverter, and a charge controller. P.V. panels are the primary component, capturing sunlight and converting it into electrical energy. The role of the inverter is to convert the direct current (D.C.) produced by the panels into alternating current (A.C.), which is used by most household appliances and tools.

Batteries store this generated power, providing energy during nighttime or overcast conditions when solar panels cannot produce electricity. The lifetime and safety of the system are guaranteed by the charge controller, which controls the flow of power from the panels to the battery and guards against overcharging and damage.

The scalability and portability of solar installations make them remarkably adaptable for emergencies. Systems can be designed to fit specific needs, from small portable kits that power individual homes or field operations to larger, more permanent installations that can energize relief centers or hospitals. Portable solar generators are especially valuable in emergencies; they can be quickly deployed to various locations, providing immediate power without needing permanent installation.

These mobile units are essential for field hospitals, temporary shelters, and rescue operations, offering a flexible and rapid solution to power needs in critical situations. For instance, after Hurricane Maria devastated Puerto Rico in 2017, solar power played a crucial role in the island's recovery. With the electrical grid almost destroyed, solar companies donated portable solar panels and batteries that were instrumental in restoring power to homes and emergency services.

Hospitals equipped with solar power were among the first to regain electrical service, which was vital for treating patients and preserving medications and vaccines in cold storage.

In another case, where wildfires and subsequent power shutoffs have become increasingly common in California, many households and emergency services have turned to solar power coupled with battery storage to maintain electricity during fire seasons. These systems ensure that communication lines remain open and evacuation orders can be disseminated promptly, significantly enhancing public safety.

Implementing solar energy systems for emergency preparedness also involves careful planning and consideration of local conditions. It's important to assess the area's typical weather patterns and potential disaster risks to design a system that can operate effectively under those conditions. For instance, areas prone to heavy cloud cover or shorter daylight hours might require larger battery storage capacity or additional P.V. panels to capture sufficient solar energy.

Moreover, community-level strategies can amplify the benefits of solar power in emergencies. Establishing community charging stations and solar-powered emergency shelters can ensure broader access to electricity during disasters, enhancing resilience and support for vulnerable populations.

Integrating solar power into emergency preparedness plans provides a dependable, sustainable, and flexible energy solution that enhances resilience and ensures the continuity of critical functions during disasters. As technology advances and the cost of solar components continues to decrease, the role of solar energy in emergency and disaster preparedness is expected to grow, reinforcing its importance as a critical component of modern, resilient infrastructure.

Off-grid solar for Remote Locations

Implementing off-grid solar systems in remote locations presents unique challenges that require specialized solutions to ensure reliable and consistent power supply. These areas, often isolated from conventional energy grids, depend heavily on self-sustaining systems. Solar power offers a compelling solution with its capacity for decentralization and sustainability. However, understanding the specific considerations for system sizing, energy storage, and weather adaptability is crucial for successful implementation.

One of the primary challenges in remote locations is accurately sizing the solar power system to meet the energy needs of the application, whether it's a residential home, a research station, or a community facility. System sizing involves calculating the total electrical load and understanding the usage patterns to ensure the solar system can produce enough power throughout the year.

Key to this process is an assessment of the electrical appliances and equipment that will be powered by the system, quantifying their energy consumption in watts and considering the hours of operation per day. This data, combined with local solar insolation values—which measure the amount of solar radiation received in a particular area—helps determine the solar panel array's necessary capacity and the inverter's size.

Energy storage is another critical component of off-grid solar systems, especially in remote areas where reliability is paramount. Batteries are vital in storing surplus solar energy produced during peak sunlight hours for use during nighttime or cloudy days. The choice of batteries—lead-acid, lithium-ion, or others—depends on factors such as lifecycle, maintenance requirements, temperature sensitivity, and budget. Advanced battery management systems are also integral to optimizing battery performance and extending lifespan by preventing overcharging and deep discharge.

Weather conditions in remote areas can significantly affect the performance of solar power systems. In regions with high variability in weather, such as extended periods of rain or snow, solar systems must be designed to maximize light absorption and minimize losses.

This might involve installing larger solar arrays to compensate for reduced solar gains during overcast days or incorporating tracking systems that adjust the angle of the panels to follow the sun, thereby increasing their efficiency. Additionally, the physical construction of the panels must be robust enough to withstand harsh weather conditions, including high winds, snow loads, and potential impacts from debris.

Maintaining a consistent power supply in varying weather conditions also involves integrating alternative energy sources to create a hybrid system. For example, pairing solar power with wind turbines or diesel generators can provide an additional layer of energy security. Such hybrid systems are beneficial in regions with seasonal or insufficient solar potential to meet all energy needs throughout the year.

Implementing off-grid solar systems in remote locations also requires careful planning and consideration of the logistical challenges of transporting and installing equipment in inaccessible areas. The use of modular, easily transportable components can simplify installation and maintenance. Remote monitoring technologies enable real-time system performance tracking and troubleshooting, reducing the need for technicians to make frequent on-site visits.

Moreover, training residents or users in essential system maintenance can enhance the reliability of the solar installation. Empowering local communities with the knowledge to perform routine checks and minor repairs can lead to more sustainable and long-lasting energy solutions.

Lastly, regulatory and financial considerations can influence the feasibility of deploying off-grid solar systems in remote areas. Incentives such as grants, subsidies, or tax breaks can offset the initial high costs of setting up off-grid solar systems and encourage widespread adoption.

The implementation of off-grid solar systems in remote locations comes with significant challenges, but advancements in solar technology, battery solutions, and system design innovations continue to improve their feasibility and efficiency. These systems provide a reliable power source and contribute to remote communities' sustainability and self-sufficiency, ultimately making a crucial impact on their development and quality of life.

CHAPTER 14

NAVIGATING LEGAL AND REGULATORY CHALLENGES

Navigating the legal and regulatory landscape is a critical aspect of implementing solar power projects, where adherence to a myriad of rules and standards is essential for successful deployment. As solar energy continues to grow in popularity, understanding and complying with these regulations becomes increasingly important, not only to ensure the legality of the installation but also to maximize available incentives. These challenges encompass a range of local, state, and federal regulations, which can include zoning laws, building codes, and specific solar access rights.

Additionally, the dynamic nature of these legal frameworks means that they can frequently change, requiring ongoing attention and adaptation from project developers and homeowners alike. Effective navigation through this complex regulatory environment is crucial for securing project approval, optimizing system performance, and ultimately contributing to the broader integration of solar energy into the national grid. This complex backdrop sets the stage for a detailed exploration of how solar power installations can meet regulatory requirements and leverage legal frameworks to enhance sustainability and economic viability.

Understanding Solar Power Regulations

Solar power regulations encompass a broad array of federal, state, and local directives that shape the development, installation, and operation of solar energy systems. These regulations are designed to ensure safety, promote efficiency, and support the integration of solar energy into the existing energy infrastructure. Understanding these regulatory layers is crucial for anyone planning, deploying, or managing solar power projects.

At the federal level, solar power regulations primarily focus on safety standards and national incentives. Central to the development of these regulations are the U.S. Department of Energy (DOE) and the Environmental Protection Agency (EPA). Key federal regulations include the National Electric Code (NEC), which sets standards for the safe installation of electrical wiring and equipment, including solar panels and related devices.

Compliance with the NEC helps prevent electrical fires, shocks, and other hazards. Furthermore, the federal government offers financial incentives such as the Solar Investment Financial Credit (ITC), which has played a significant role in the growth of the solar industry. Solar projects become more financially viable when households and companies may deduct a percentage of their solar expenditures from their taxes thanks to the Investment Tax Credit (ITC).

State regulations can vary widely, reflecting the diverse policies and priorities of individual states. Renewable-sourced electricity must meet certain requirements by a certain date, as per state-implemented renewable portfolio standards (RPS). These standards are crucial in driving the adoption of solar power at the state level by creating a guaranteed market for renewable energy.

Additionally, states may offer various incentives such as rebates, tax breaks, and grants to encourage the adoption of solar technology. For example, some states provide net metering policies, allowing solar system owners to sell excess power back to the grid at retail prices, effectively reducing their utility bills.

At the local level, solar power regulations typically involve zoning and building codes, permits, and sometimes solar access laws that protect a property owner's right to sunlight. Local governments may require solar panel projects to receive specific permits before installation can commence. These permits ensure that solar installations do not violate local zoning laws, which might regulate where panels can be placed and how they should be installed to maintain community aesthetic standards and safety.

For instance, some localities have regulations that prevent the installation of panels in front yards or require that they be set back a certain distance from the property line.

Moreover, solar access laws at the local or state level can protect homeowners from having their solar panels shaded by future construction or tree growth on neighboring properties. These laws are vital for ensuring that investments in solar technology are safeguarded against actions that could diminish their effectiveness.

Challenges in navigating the regulatory landscape for solar power often involve the complexity and variability of regulations from one jurisdiction to another. Solar installers and project developers must be well-versed in the specific requirements of the areas where they operate to avoid legal conflicts and delays. This might involve engaging with local officials, participating in public hearings, and staying informed about changes in legislation.

Education and advocacy are also important components of dealing with solar power regulations. Industry associations and nonprofit organizations often work to educate the public and policymakers about the benefits of solar power and advocate for regulatory changes that support the growth of solar energy. These entities play a crucial role in pushing for regulatory frameworks that facilitate rather than hinder the adoption of solar technology.

The regulatory landscape for solar power is complex and multi-layered, involving a range of guidelines and standards at the federal, state, and local levels. Navigating these regulations requires a thorough understanding of applicable laws and incentives, as well as proactive engagement with regulatory processes.

Successfully managing these regulatory challenges not only ensures compliance but also maximizes the potential benefits of solar projects, contributing to the broader adoption and sustainability of solar energy. As the solar industry continues to evolve, staying informed and adaptable to regulatory changes will remain a key component of success in this dynamic field.

Navigating Zoning Laws and Permits

Achieving the required permissions and navigating zoning regulations are essential elements in installing solar power systems. This process involves understanding local regulations, interacting effectively with local authorities, and overcoming various bureaucratic obstacles. An outline of the actions necessary to effectively handle these obstacles and guarantee a seamless installation procedure is given in this guide.

Understanding Local Zoning Laws

The first step in navigating zoning laws is to gain a thorough understanding of the local regulations that apply to solar installations. Zoning laws vary significantly from one jurisdiction to another and can affect various aspects of a solar project, including where panels can be installed on a property, how they must be mounted, and even the types of solar systems allowed.

Homeowners should start by consulting their local building department or zoning office to gather information on specific solar zoning regulations. Many municipalities also have online resources that provide guidelines on local solar installations.

It's important to determine if the area is zoned for the type of solar installation planned. Some areas may have restrictions based on historic preservation, aesthetic considerations, or building density. In such cases, special approvals or variances may be required. Understanding these nuances is crucial for planning the installation in compliance with local norms and avoiding costly adjustments later.

Obtaining Necessary Permits

Securing the necessary permits is a multi-step process that usually involves several stages of approval. The typical permitting process for a solar installation includes submitting detailed plans of the system to the local building department. These plans should include specifications of the solar panels, the mounting system, wiring, inverters, and any other components that form part of the installation.

Permit applications must often be accompanied by a site plan that shows where the solar panels will be located on the property and how they will be integrated into existing structures. The reviewing authority will assess the plans to ensure they comply with local building codes, zoning laws, and safety regulations. This review process can take from a few days to several months, depending on the complexity of the project and the efficiency of the local offices.

Dealing with Local Authorities

Effective communication with local authorities is key to navigating the permitting process smoothly. It is beneficial to establish a good working relationship with the relevant officials and to approach the process with a mindset of collaboration. Be prepared to answer questions about the project and to provide additional documentation as needed.

It's also helpful to attend any required public hearings or community meetings. These can be opportunities to explain the benefits of the project to the community and address any concerns that neighbors or local officials might have. Demonstrating a commitment to adhering to local regulations and being responsive to community input can facilitate smoother approval processes.

Common Obstacles

Common obstacles in the permitting process can include delays in review, requests for additional information, or outright denials based on non-compliance with specific codes. To overcome these obstacles, it's crucial to:

Ensure complete and accurate application materials: Incomplete applications are a common reason for delay. Double-check that all required documents are included before submission.

Be proactive about compliance: Understand local codes thoroughly and design the project to comply from the outset. This reduces the likelihood of rejection due to non-compliance.

Plan for contingencies: In some cases, local laws may change during the planning or installation process. Staying informed about potential changes and being prepared to adjust plans accordingly can help mitigate these risks.

Seek professional help: Hiring a professional who has experience with local solar installations can be invaluable. These professionals can offer insights into the local regulatory landscape and may have established relationships with the local authorities, facilitating a smoother permitting process.

Insurance Considerations for Solar Installations

Insurance for solar installations is a critical aspect of safeguarding an investment in renewable energy. As solar power continues to gain popularity, both residential and commercial property owners must understand the specific insurance considerations that apply to their solar panels and associated equipment. This understanding helps manage risks effectively and ensure that solar installation remains a viable and protected investment over the long term.

When considering insurance for solar installations, it is essential to start with property insurance. This type of insurance generally covers the physical damage to or loss of the solar panels themselves, along with any associated equipment, such as inverters and battery storage systems. Damage could arise from a variety of sources, including environmental events like hail, storms, or heavy snowfall, as well as human-related incidents such as vandalism or theft. Property insurance policies need to be carefully reviewed to ensure that they explicitly include solar panels as covered items. In some cases, solar installations might increase the value of the property significantly, which may require adjustments to the overall coverage limits of an existing property insurance policy.

Liability insurance is another crucial consideration. This type of insurance protects the solar system owner from legal and medical costs that could arise if the solar installation causes harm to a third person or damage to property. For example, if a poorly secured panel were to detach and cause injury or damage during a storm, liability insurance would cover the associated costs, protecting the owner from potential financial loss.

In commercial settings, where solar installations are often larger and more complex, additional types of insurance may be necessary. Business interruption insurance, for example, can be vital for operations that rely heavily on their solar systems for daily functioning.

If a solar system fails or is damaged and causes a halt or reduction in business operations, business interruption insurance can compensate for lost income during the period of system downtime. This coverage is particularly important for businesses in sectors like manufacturing or data centers where consistent power supply is critical.

Performance guarantees and warranties provided by solar equipment manufacturers or installers are also an integral part of the insurance landscape. These guarantees typically assure a certain level of operational efficiency and performance over a defined period. To complement these warranties, some solar project owners opt for specialized warranty management insurance products that cover the cost of repairs or replacements if the original manufacturer fails to honor their warranty commitments, possibly due to bankruptcy or other financial issues.

When setting up insurance for a solar installation, it is advisable to work with insurance companies that have experience with renewable energy projects. These insurers will be more knowledgeable about the particular dangers connected to solar systems and will be able to provide solutions that are designed to fulfill these demands. They can also offer advice on risk-management techniques that will further safeguard the solar investment.

These practices might include regular maintenance schedules, the installation of protective measures like lightning arresters, and the use of high-quality, durable materials that can withstand environmental stresses.

Engaging in a detailed discussion with an insurance agent to explore all possible risks and coverage options is a critical step. During these discussions, it is important to clarify aspects such as the insurance company's claims process, the extent of deductible amounts, and the procedures for updating the policy as the solar installation expands or upgrades over time.

Furthermore, solar system owners should consider the potential future changes in insurance requirements that could arise from evolving local regulations or changes in technology. Staying informed about these developments can help ensure that the solar installation remains adequately protected throughout its operational lifetime.

Leveraging Government Incentives and Subsidies

Solar power has emerged as a key component of global renewable energy strategies, driven in part by significant government incentives and subsidies designed to promote its adoption. These incentives and subsidies aim to make solar power more accessible and financially attractive for both individuals and businesses. Understanding the types of incentives available, as well as how to access and qualify for them, is crucial for anyone considering investing in solar energy.

Government incentives for solar power typically include tax credits, rebates, and grants, which can substantially reduce the initial cost of installation and even improve the return on investment over time. One of the most significant incentives in the United States has been the Federal Investment Tax Credit (ITC), which allows homeowners and businesses to deduct a substantial percentage of their solar installation costs from their federal taxes.

Originally set at 30%, the ITC has undergone periodic adjustments and extensions, reflecting ongoing legislative support for solar energy. This tax credit directly reduces the amount of tax owed rather than just reducing taxable income, making it a powerful incentive for the adoption of solar technology.

On top of federal tax credits, many states offer additional incentives. These can include state-specific tax credits, property tax exemptions, and sales tax exemptions. For instance, some states allow a property tax exemption on the added home value due to the installation of solar panels. This means that while the solar installation may increase the value of a property, the owner's property taxes will not increase as a result. Sales tax exemptions can further reduce the cost of solar equipment by eliminating the state sales tax at the point of purchase.

Rebates are another form of incentive that can lower the upfront cost of solar systems. Provided either by state governments, local municipalities, or even utility companies, rebates are typically available for a limited time or until funding is exhausted. These rebates can be straightforward cash offers that reduce the purchase price, making solar installations more financially accessible.

Additionally, some regions offer performance-based incentives (PBIs), which provide monetary rewards based on the electricity that a solar system produces over a certain period. Unlike upfront incentives, PBIs reward the actual performance and continued operation of the system, typically on a per-kilowatt-hour basis. This not only helps offset the initial costs but also encourages optimal maintenance and operation of solar installations.

To access these incentives, solar system purchasers must navigate a variety of application processes and eligibility criteria. The first step is to conduct thorough research or consult with a professional to understand what specific incentives are available in their area. This research should include visits to official government or utility websites or consultations with local solar energy associations.

Once potential incentives are identified, the next step involves understanding the eligibility requirements for each. These might include conditions related to the type of solar equipment, the qualifications of the installer, or the location of the installation. For federal tax credits, for instance, eligibility is generally contingent on owning the solar system (as opposed to leasing it) and having sufficient tax liability against which to claim the credit.

The actual application process for these incentives usually requires detailed documentation. For tax credits, this might involve keeping receipts and proof of installation. For rebates or grants, applicants may need to submit detailed proposals or installation plans. Working with an experienced installer who is familiar with local incentive programs can streamline this process, ensuring that all requirements are met and maximizing the financial benefits.

Lastly, staying informed about changes in government policies is critical, as incentives can vary annually based on budget allocations, legislative changes, or shifts in governmental energy strategies. Engaging with local solar advocacy groups and regularly reviewing official energy resource sites can provide updates and insights into the evolving landscape of solar incentives.

Staying Compliant with Electrical Codes

Compliance with electrical codes is not merely a bureaucratic step in the process of installing solar power systems; it is a critical safety measure that ensures these systems operate reliably and safely over their intended lifespan. Electrical codes set the minimum standards for electrical system safety to protect people, property, and the environment from electrical hazards. These regulations are developed based on thorough research, past incidents, and ongoing technological advancements. They address everything from wire size and panel capacity to system grounding and proper circuit protection.

Understanding and adhering to these codes is essential for several reasons. Firstly, compliance helps prevent the risk of electrical fires, which can occur due to poor wiring, overloads, or faulty installation. It also guards against electrical shocks and other safety hazards that could endanger both installers and users of the solar system. Moreover, proper compliance ensures that the solar power system operates efficiently and delivers the expected energy output without frequent disruptions or failures.

Common Compliance Issues

One of the most common compliance issues in solar installations is improper wiring. This can include using incorrect wire types or sizes, inadequate insulation, or faulty connections that may lead to overheating, reduced system performance, or even fires. To address this issue, it is crucial to follow the National Electrical Code (NEC) or other local codes that specify the requirements for wiring solar power systems, including the minimum wire gauge, type of insulation, and protection against environmental factors like UV light and moisture.

Another frequent challenge is the incorrect installation of inverters and electrical panels. Inverters must be installed in locations that adhere to clearance regulations to ensure they remain accessible and maintain proper ventilation. The electrical panels, similarly, need to be configured correctly to handle the load and distribution of solar-generated electricity safely. Compliance issues here can lead to system inefficiencies or hazards, especially if the system's electrical load exceeds the panel's capacity.

Grounding issues are also prevalent, where either inadequate grounding or a total lack of a proper grounding system can expose the installation to potential lightning strikes or power surges, posing serious safety risks. The grounding system needs to meet specific resistance requirements and should be installed in a manner that effectively protects both the system components and the occupants of the building.

Addressing Compliance Issues

To effectively address these and other compliance issues, the first step is a thorough understanding of the applicable electrical codes. This knowledge can typically be gained through professional training courses, updated manuals, and direct consultations with local building officials or experienced installers. Keeping abreast of changes in the NEC, which is updated every three years, is particularly important as it reflects new safety practices, technological innovations, and lessons learned from past incidents.

Involving a certified electrical inspector early in the installation process can help identify potential compliance issues before the system is fully installed. These professionals can provide preliminary reviews of the design and installation plans to ensure that all aspects of the system are up to code. Their early input can prevent costly and time-consuming rework after the installation is completed.

Documentation plays a pivotal role in ensuring and proving compliance. This includes keeping detailed records of all design calculations, equipment specifications, and installation procedures. Such documentation is not only useful during inspections but also serves as a reference throughout the lifetime of the solar installation, particularly during maintenance or troubleshooting.

For complex installations, using specialized software for design and simulation can help visualize potential problems and test scenarios under different conditions to ensure they meet code requirements. This software can model electrical loads, assess circuit protection needs, and simulate system responses to faults, providing an in-depth analysis that supports compliance.

Finally, ongoing education and training for anyone involved in designing, installing, or maintaining solar power systems are crucial. As technologies and standards evolve, continuous learning ensures that professionals are equipped to apply the most current and effective practices in their work.

CHAPTER 15

THE FUTURE OF SOLAR TECHNOLOGY

As we look toward the horizon of renewable energy technologies, solar power stands at the forefront of innovation and potential. The future of solar technology promises significant advancements that could revolutionize how we harvest and utilize solar energy. Researchers and engineers around the world are focusing on enhancing the efficiency of photovoltaic cells, developing new materials that could lead to cheaper and more flexible solar panels, and integrating cutting-edge technologies like artificial intelligence to optimize energy production and grid management.

Additionally, the push for more sustainable energy solutions is driving the exploration of novel solar applications, from transparent solar panels that can coat windows and buildings to solar fabrics that can power devices on the go. This chapter delves into these emerging technologies and explores how they are set to transform the landscape of renewable energy, making solar power more accessible, efficient, and integral to our daily lives.

Emerging Trends in Solar Technology

The solar technology landscape is rapidly evolving, driven by both advancements in photovoltaic materials and the growing need for more efficient, durable, and versatile solar solutions. These innovations not only enhance the effectiveness of solar panels but also broaden their applicability across various sectors, contributing significantly to the global shift towards renewable energy.

One of the most notable trends in solar technology is the development of new photovoltaic materials that offer greater efficiency and lower production costs than traditional silicon-based cells. Perovskite solar cells, for instance, have emerged as a promising alternative due to their high efficiency and versatility.

These materials can be manufactured using simpler processes than those required for silicon cells, potentially reducing costs and increasing the adaptability of solar panels to different environments. Perovskites can be used to create thin-film solar cells that are lightweight and flexible, allowing them to be installed on surfaces where traditional panels would be impractical.

Another significant advancement is in the area of solar efficiency improvements. Researchers are continuously finding ways to increase the percentage of sunlight that solar panels can convert into usable energy. Techniques such as using multi-junction cells, which layer several types of solar cells atop one another to capture different segments of the solar spectrum, have shown the potential to achieve efficiency rates far beyond the current average.

Additionally, incorporating nanotechnology into solar panel design can help in developing ultra-thin solar cells that capture sunlight more effectively and with minimal loss of energy.

Innovations in solar energy storage also represent a critical area of development. As the adoption of solar energy grows, the need for effective storage solutions becomes increasingly important, particularly for managing supply and demand fluctuations and ensuring a steady energy supply during nighttime or overcast conditions. Lithium-ion batteries have dominated the solar storage market due to their high efficiency and capacity. However, new technologies such as flow batteries, which store energy in liquid chemical solutions, and solid-state batteries, which use solid electrolytes and offer higher safety and density, are beginning to gain traction. These technologies promise longer lifespans and higher capacities compared to traditional battery systems, making them particularly suitable for large-scale energy storage.

Additionally, the integration of solar power with smart grid technology is an emerging trend that promises to revolutionize how energy is managed and distributed. Smart grids use digital communication technology to control the flow of electricity from suppliers to consumers.

Integrating solar systems with smart grids allows for better energy management, enabling utilities to effectively handle the variable nature of solar power and optimize grid stability. Additionally, by enabling the growth of decentralized energy systems—which generate and use energy locally and so lessen dependency on massive infrastructure—integration also contributes to increased energy security.

Furthermore, the push for sustainability is driving the trend towards recycling and repurposing materials used in solar panels. The necessity for the solar industry to handle end-of-life panels in an ecologically appropriate way is growing as the number of solar installations worldwide rises. Recycling technologies that can recover valuable materials like silver, silicon, and glass from old solar panels are becoming more refined and widely implemented, promoting a circular economy within the solar industry.

Innovations in Solar Panel Materials

Recent innovations in solar panel materials are transforming the landscape of solar technology, enhancing both the performance and durability of solar panels. The efficiency with which solar panels convert sunshine into electricity is being enhanced by these innovations, and they are also lowering costs, which makes solar power a more viable and accessible primary energy source.

The use of perovskite materials in solar panel manufacturing is one of the most promising advances. Perovskites are a class of materials that have shown significant potential to offer high efficiency at low production costs. Unlike traditional silicon-based solar cells, perovskite solar cells can be manufactured using simpler liquid-based processes, potentially reducing manufacturing costs and energy consumption.

They can also be made semi-transparent and integrated into windows or other surfaces, opening new avenues for solar energy harvesting in urban environments. Recent research has pushed the efficiency of perovskite solar cells close to that of silicon cells, achieving rates of over 25%. However, the primary challenge remains the long-term stability of perovskite materials, as they are more prone to degradation when exposed to moisture and UV light.

Another innovative material is gallium arsenide, which has been used in the design of multi-junction solar cells. These cells are made up of several layers made of various materials, each of which is intended to capture a distinct region of the sun spectrum. This arrangement significantly increases the overall efficiency of the solar panels, with some laboratory examples reaching efficiencies above 40%. Although currently expensive, gallium arsenide solar cells are being used in specialized applications, such as satellites and space probes, where high efficiency in a compact form is crucial. As manufacturing processes improve and scale up, it's expected that the costs associated with these high-efficiency cells will decrease, making them more viable for broader applications.

Graphene is another material that is finding its way into solar panel technology. Known for its extraordinary electrical and thermal conductivity, strength, and flexibility, graphene can be used to create transparent conducting films for solar panels that are more efficient and less expensive than those made from traditional materials like indium tin oxide. The incorporation of graphene layers into solar cells can help improve their efficiency by reducing energy loss that typically occurs at the metal-semiconductor interface.

Moreover, graphene's flexibility allows for the development of bendable solar panels, which can be used on surfaces that are not flat, opening up a wide range of new applications. In addition to these materials, advancements are also being made in the coatings and encapsulants used with solar panels. For instance, new anti-reflective coatings and textured surfaces are being developed to reduce the amount of light that is reflected away from the solar cell, thereby increasing the amount of light absorbed and improving the efficiency of the solar panels.

Enhanced encapsulant materials that provide better protection against environmental factors like moisture and UV radiation are also being developed to extend the lifespan of solar panels. The impact of these innovations on the cost and efficiency of solar panels is profound. By improving the efficiency of solar panels, less surface area is needed to generate the same amount of power, reducing the amount of material required per watt of power produced. This not only lowers the cost of the solar panels themselves but also reduces the associated installation and material costs.

Moreover, as these new materials and technologies become more widely used and manufacturing processes mature, economies of scale are expected to bring costs down further, making solar energy an even more competitive alternative to traditional fossil fuels. Overall, the continued innovation in solar panel materials is critical for advancing the field of solar energy. With each breakthrough, solar power becomes more efficient, more durable, and more cost-effective, paving the way for its increased use in meeting the world's energy demands sustainably.

The Role of Artificial Intelligence in Solar Power

By improving the efficiency, dependability, and integration of solar power systems, artificial intelligence (AI) is completely changing the solar sector. AI technologies are being used in many areas of solar energy, from grid integration and maintenance to production and monitoring. These applications offer creative solutions that raise the efficiency and profitability of solar systems.

Optimizing Energy Production

AI is particularly effective in optimizing the energy production of solar panels. By analyzing vast amounts of data from weather forecasts, historical weather patterns, and real-time solar irradiance, AI algorithms can predict solar output with high accuracy. This predictive capability allows for more precise adjustments to the positioning of solar panels throughout the day, maximizing their exposure to sunlight and thus optimizing energy production. Furthermore, AI can dynamically adjust energy output based on current demand and grid conditions, ensuring that solar power generation is as efficient as possible.

Predictive Maintenance

The field of maintenance benefits greatly from the integration of AI in solar power systems. Due of the unanticipated downtime they frequently cause, traditional reactive maintenance techniques can be expensive and ineffective. AI introduces a predictive maintenance approach, where algorithms analyze data from solar panels to predict failures before they occur.

By monitoring parameters such as voltage outputs, temperature profiles, and even the visual appearance of solar cells through image recognition technologies, AI can identify anomalies that may indicate potential issues. By taking early measures to resolve these problems, maintenance crews may reduce downtime and increase the solar panels' lifespan.

Grid Integration

The incorporation of solar electricity into the electrical grid is another important function of AI. Solar energy's variable nature can pose challenges to grid stability. AI systems help mitigate these challenges by intelligently managing the energy flow from solar panels to the grid. Through real-time data analysis and learning algorithms, AI can predict peak demand periods and adjust the supply accordingly. AI can also make it easier for companies and homes to trade excess solar energy they create with other members of the grid network, maximizing the usage of solar power in local microgrids. This is known as distributed energy trading on a distributed grid.

Advanced Monitoring and Control Systems

Advanced monitoring and control systems powered by AI further enhance the capabilities of solar power systems. Operators may gain comprehensive insights into the operation of their solar installations with the help of these systems, which continually gather and analyze data on solar production and consumption trends. This information can be used to make informed decisions on system adjustments, investment in additional capacity, and even policy development for energy management.

Challenges and Future Directions

While the integration of AI in the solar industry offers significant benefits, there are challenges that need to be addressed. Since AI systems need access to vast amounts of data, some of which may be sensitive, data privacy and security are important problems. Furthermore, smaller operators and home users may find it prohibitive to use AI technology due to their hefty upfront costs.

Despite these challenges, the future of AI in the solar industry looks promising. Continued advancements in AI technology and data analytics are expected to drive further improvements in solar power efficiency and grid integration. AI is expected to have an even more dramatic effect on the solar industry as the technology becomes more accessible and affordable, making solar energy a more viable option for the global energy mix. This continuous development highlights the solar industry's dynamic character and its ability to support global sustainable energy solutions.

The Impact of Energy Storage Advancements

The solar power sector is undergoing a revolution because of developments in energy storage technology, which also improve the clean energy source's accessibility and dependability. One of the main obstacles to the widespread use of solar power is its intermittency, which may be addressed by incorporating better storage systems. Energy storage systems enable solar power to transcend limitations related to weather and daylight variability, making solar energy a more consistent and dependable energy source.

Energy storage technology, particularly in the form of batteries, has seen significant advancements in efficiency, capacity, and cost-effectiveness. Lithium-ion batteries, which have become the standard in the renewable energy sector, are renowned for their high energy density and rapid charging capabilities.

These batteries have undergone continuous improvements that have increased their lifespan and decreased their susceptibility to temperature variations, making them more suitable for a range of climates and applications. Furthermore, the cost of lithium-ion batteries has plummeted in the past decade, driven by scaling production and technological improvements, thereby reducing the overall cost of solar energy storage systems.

Beyond lithium-ion, other battery technologies are emerging that promise even greater improvements. Solid-state batteries, for example, replace the liquid or gel electrolytes found in conventional batteries with solid materials. This change can potentially offer higher energy density, enhanced safety, and longer lifespans. These batteries are less prone to risks like electrolyte leakage or fire, which can further increase their appeal in residential and commercial applications. Solid-state batteries offer a substantial advancement in energy storage technology, even if they are still in the early phases of research.

Another promising development in the field of energy storage is the advent of flow batteries. These systems store chemical energy in liquid solutions in external tanks rather than within the battery itself. This design allows for quick energy release and scalability in a way that traditional batteries cannot match.

Flow batteries are particularly well-suited for storing large amounts of energy, which makes them ideal for use in utility-scale solar installations. By supplying electricity at times of high demand, they can aid in grid stabilization and improve the integration of solar power into already-existing power networks.

The impact of these advancements extends beyond merely enhancing solar power's reliability; they also make solar energy more accessible. With more efficient and affordable storage solutions, solar power can be deployed in more regions, including remote and underserved areas where grid connection is not feasible. This accessibility has the potential to significantly alter the energy landscape, especially in poor nations where a large number of populations still rely on costly and environmentally harmful diesel generators for electricity.

For the first time, these communities will have access to clean, dependable, and reasonably priced electricity thanks to the capacity to store solar energy effectively. Moreover, the integration of advanced energy storage with solar systems allows for greater flexibility in energy management within the grid.

For example, energy storage can provide load leveling, which balances the supply and demand of electricity throughout the day by preventing the peaks and troughs that can lead to energy wastage or necessitate the use of peaking power plants, which often run on fossil fuels. Additionally, during periods of excess generation, energy can be stored instead of being wasted or sold off at negative prices.

Moreover, energy storage is essential for improving the power grid's resiliency. By providing a reliable backup power source, energy storage systems can help maintain electricity supply during grid outages caused by natural disasters or system failures. This resilience is especially crucial at a time when climate change is making extreme weather occurrences more common.

Predicting the Future of Renewable Energy

With solar power positioned to play a key part in the worldwide transition to a cleaner and more sustainable energy landscape, the future of renewable energy appears increasingly promising. As technological advancements continue and policy frameworks evolve, solar energy is expected to become even more prevalent, driven by both economic and environmental considerations.

Factors Influencing the Growth of Solar Energy

Several key factors are set to influence the growth of solar energy in the coming years. First, technological advancements are continually improving the efficiency and reducing the cost of solar panels. Technological advancements like perovskite solar cells and bi-facial panels—which have the ability to absorb light from both sides—are advancing solar efficiency. Solar energy will probably continue to get cheaper as these technologies advance and become widely used, giving it a stronger competitive edge over conventional fossil fuels.

The development of the solar business is greatly aided by government initiatives and laws. Numerous nations globally have established audacious goals for renewable energy and are bolstering them with inducements, including tax credits, subsidies, and feed-in tariffs. For instance, the European Union's Green Deal and the United States' renewable energy subsidies under various legislative acts aim to significantly boost the capacity of solar installations.

As climate change mitigation climbs higher on the global agenda, such policy support is expected to strengthen, providing a robust framework for the expansion of solar energy.

Market dynamics related to energy consumption and the increasing demand for electricity globally also support the expansion of solar power. With rising energy needs in developing economies and the global push towards electrification, including transportation and heating, solar energy offers a scalable and rapidly deployable solution to meet this demand sustainably.

Challenges to Growth

Despite the positive outlook, the growth of solar energy faces several challenges that could impact its trajectory. One of the primary obstacles is the intermittent nature of solar power, which results from its reliance on sunshine and makes it unavailable in all -weather situations or times of the day. This variability poses a significant challenge for grid integration and reliability. Advances in energy storage technology, such as more efficient batteries or new concepts like hydrogen storage, are critical to mitigating this issue and making solar power a reliable part of the energy mix.

Infrastructure and regulatory challenges also pose significant barriers. The integration of large-scale solar power into the existing grid requires substantial upgrades to infrastructure and regulatory frameworks that can handle the decentralized and variable nature of renewable energy sources. Additionally, as solar power becomes a larger part of the energy mix, changes in how energy markets operate and are regulated will be required.

Environmental and social impacts, although significantly less than those associated with fossil fuels, also need consideration. Waste is produced, and hazardous chemicals are used to manufacture solar panels. Responsibly managing these materials and advancing recycling technologies will be crucial to minimizing the environmental footprint of solar power. Furthermore, large-scale solar installations require significant land areas, which could conflict with agricultural uses or biodiversity conservation efforts.

The Path Forward

Looking ahead, the trajectory of solar power is one of exponential growth and central importance in achieving global energy transition goals. Continued innovation in technology and materials science, supportive policy frameworks, and advancements in grid infrastructure and energy storage will be pivotal in overcoming challenges and realizing the potential of solar energy.

As stakeholders, from policymakers to private investors, focus on these areas, the future of solar energy promises not only enhanced sustainability but also greater energy security and economic benefits globally. The next few decades are likely to see solar power move from a supplementary source of energy to a cornerstone of the global energy system, playing a vital role in combating climate change and leading the renewable energy revolution.

How You Can Share Your Review

Through Amazon.com

- Visit the Amazon page where you purchased or found my book.
- Scroll down to the 'Customer Reviews' section near the bottom of the page.
- Click on 'Write a customer review' to begin sharing your valuable insights and experiences with the book.

Instant QR Code Access

Simply scan the QR code below with your smartphone to be directly taken to the Amazon review section for the book. This quick access method makes it easy for you to leave your feedback without navigating through the website.

Your review is not just feedback for us; it's a beacon for future readers navigating the world of solar energy and sustainable living. Thank you for taking the time to share your thoughts and helping us spread the message of sustainability.

Your Review

Your review matters to us for several reasons. Firstly, it provides us with valuable feedback, helping us understand what we did well and where we can improve. This insight is crucial for enhancing the quality of our work and ensuring that it meets your needs and expectations. Secondly, your review helps to inform potential readers about the value and relevance of our book, guiding their decision to explore the topics we've discussed.

Reviews can highlight the practical benefits, clarity of explanation, and the overall impact the book might have on someone's understanding of solar energy and sustainable living. Additionally, your engagement through reviews fosters a sense of community and shared learning among readers with similar interests. It encourages dialogue and the exchange of ideas, which are fundamental to spreading awareness and adoption of sustainable practices.

Lastly, positive reviews can significantly contribute to the success and visibility of the book, enabling us to reach a wider audience and make a more substantial impact on promoting sustainable energy solutions. Your feedback is not just appreciated; it's integral to our mission of educating and inspiring action towards a more sustainable future.

Thanks Readers

Thank you for embarking on this journey through the pages of our book. Your interest and dedication to exploring the world of solar energy and sustainable living are what inspire us to share our knowledge and insights. We hope that this guide has provided you with valuable information, practical advice, and the inspiration needed to harness the power of the sun for a more sustainable and energy-independent future. Whether you are taking your first steps toward solar energy or looking to expand your existing knowledge, we are grateful for your commitment to making a positive impact on our planet. Thank you for choosing our book as your guide, and we wish you success in all your solar energy endeavors.

www.ingramcontent.com/pod-product-compliance
Lightning Source LLC
Chambersburg PA
CBHW062313220526
45479CB00004B/1147

9798332120428